T0203248

Network Performance Analysis

Network Performance Analysis

Thomas Bonald
Mathieu Feuillet

Series Editor
Pierre-Noël Favennec

Library of Congress Cataloging-in-Publication Data

Bonald, Thomas.
  Network performance analysis / Thomas Bonald, Mathieu Feuillet.
    p. cm.
  Includes bibliographical references and index.
  ISBN 978-1-84821-312-8
  1. Computer networks--Evaluation. 2. Network performance (Telecommunication) 3. Queuing theory.
I. Feuillet, Mathieu. II. Title.
  TK5105.5.B656 2011
  621.382--dc23

                                                                           2011026541

British Library Cataloguing-in-Publication Data
A CIP record for this book is available from the British Library
ISBN 978-1-84821-312-8

Printed and bound in Great Britain by CPI Group (UK) Ltd., Croydon, Surrey CR0 4YY

Table of Contents

Preface

When speaking of queues, the first idea that comes to mind is that of everyday life: queues in supermarkets, airports, banks, etc. It is more difficult to imagine the queues used in computer systems and communication networks. However, these queues are crucial for smooth system operation and good performance. They are also more various and elaborate than those of everyday life just like bits and datagrams are more flexible than human beings.

Traffic being random, the analysis of queues relies on the theory of probability and more specifically on the Markov theory. This theory has a very simple principle, but a wide range of applications, and has become, during the last century, a fundamental tool for computer science and networking, but also for other scientific domains such as statistics, physics, biology, and economics. In the first four chapters of this book, we present the main results of the Markov theory, using only basic notions of probability.

The chapters dedicated to traffic and communication networks have benefited from our work experience in the laboratories of France Telecom, where we have experienced the importance of traffic modeling and performance evaluation in all the domains of network engineering: design, planning, architecture, measurement, control, etc. Analyzing each part of those huge systems that are communication networks allows us to better understand their global behavior and, *in fine*, to improve their performance.

Thomas BONALD
Mathieu FEUILLET
July 2011
Paris,
Rocquencourt

Chapter 1

Introduction

1.1. Motivation

Network performance analysis, and the underlying queueing theory, was born at the beginning of the 20th Century when two Scandinavian engineers, Erlang[1] and Engset[2], independently found very close formulas for calculating the reject probability of a telephone call. Their results have since proved instrumental in dimensioning telephone networks, to find the optimal capacity given some expected demand and target call reject rates.

Nowadays, the engineering of communication networks and computer systems, which consists of both dimensioning and designing resource-sharing algorithms and traffic control schemes, relies on mathematical tools derived from the queueing theory. The objective of this book is to describe some of these tools and to show how they are used in solving the practical engineering and performance issues.

---

1 Agner Krarup Erlang, Danish engineer and mathematician (1878–1929).
2 Tore Olaus Engset, Norwegian engineer and mathematician (1865–1943).

1.2. Networks

Roughly, there are two techniques for sharing the resources of communication networks:

– the "circuit" technique, which consists of reserving the resources prior to any communication and transferring information once the reservation is completed, along the established circuit;

– the "packet" technique, by which communications occur without any prior reservation, information being transferred in the form of independent packets subject to congestion (delay, loss) on their path to the destination.

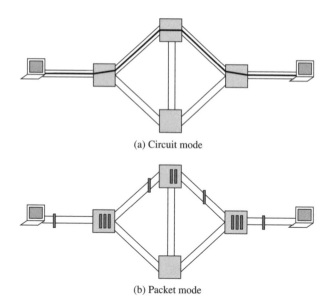

(a) Circuit mode

(b) Packet mode

Figure 1.1. *Communication techniques*

In short, this is the main difference between the (public switched) telephone network and the IP network: the principle of bandwidth reservation versus that of bandwidth sharing, the questions of accessibility (call reject rate) against those of speed (bit rate) and integrity (packet delay, packet loss rate).

The boundary between circuit mode and packet mode is not so distinct in practice. The multi-protocol label switching (MPLS) technology uses virtual circuits in IP, for instance: 3G radio access networks use both circuit and packet modes; an Internet service provider can block some video streams in case of congestion, each stream then constituting a virtual circuit in the IP network. There are many such examples. However, this broad classification between the circuit mode and the packet mode is very useful. It corresponds to two types of traffic models we shall study:

– in the circuit mode, the Erlang model and its extensions, described in Chapter 8;

– in the packet mode, the IP traffic models, described in Chapter 9 for real-time traffic (voice, video) and Chapter 10 for elastic traffic (file transfers).

1.3. Traffic

Network performance is mainly driven by the random traffic fluctuations caused by the user behavior. To find his formula in 1917, Erlang assumed, for instance, that calls arrived according to a Poisson process[3] and had exponential durations[3]. Figure 1.2(a) shows such a sequence of calls, whose durations are represented by the lengths of the horizontal bars. These assumptions allowed Erlang to apply the novel theory of Markov[3] and derive the call reject probability with respect to the number of available circuits and the traffic intensity.

Moreover, Erlang noticed that his formula was "insensitive" to the distribution of call durations beyond the mean.

---

3 We will come back to these notions in detail in Chapters 2–5.

This property, which was formally proved 40 years later[4], shows the simplicity and robustness of the Erlang formula, which depends on traffic intensity only and not on fine traffic statistics like the distribution of call durations. This also explains why the formula is still used today, though today's telephone traffic has nothing to do with that of Erlang's epoch.

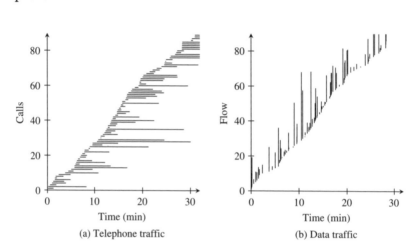

Figure 1.2. *The random nature of traffic*

Similarly, performance of the IP networks depends on the random nature of traffic. Figure 1.2(b) shows for instance data flows arriving according to a Poisson process, with exponential sizes (volumes in bytes) represented by the lengths of the vertical bars. We shall see that, under some assumptions of the way data flows share bandwidth, most performance metrics are also insensitive to the distribution of flow sizes beyond the mean. They depend on traffic statistics

---

4 B.A. Sevastyanov, *An Ergodic Theorem for Markov Processes and its Application to Telephone Systems with Refusals*, 1957.

through the traffic intensity only. These results may be viewed as the natural extensions of the Erlang formula of IP networks, with the same desirable characteristics of simplicity and robustness.

1.4. Queues

Queues are omnipresent in packet-switched networks. They are at the heart of any computer, switch, router, and access point. This is the place where sharing policies are implemented through packet scheduling and active queue management. More generally, a set of data flows sharing the same link may be viewed as a virtual queue due to the link capacity constraint, the service required by each flow corresponding to the transfer of some data volume.

By extension, the models of circuit-switched networks where calls are either admitted or rejected may be viewed as specific queues, in which customers do not wait but may be lost. Formally, we should refer to either "loss" or "waiting" queues; the simpler term of *queues* is commonly used.

1.5. Structure of the book

The book is structured as follows:

Chapter 1: Introduction;

Chapters 2–5: Poisson processes and Markov theory;

Chapters 6 and 7: Elements of queueing theory;

Chapters 8–10: Traffic models;

Chapter 11: Application to networks.

The relationship between the chapters is as follows:

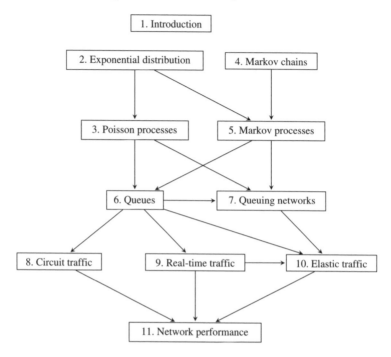

Each chapter (except for this one) contains a series of exercises with solutions. Throughout the book, we use the acronyms a.s. for "almost surely" and i.i.d. for "independent and identically distributed". We denote by $1(\cdot)$ the indicator function, by $P(\cdot)$ the probability, and by $E(\cdot)$ the expectation.

1.6. Bibliography

For further information, the interested reader is referred to the following books:

BRÉMAUD P., *Markov Chains, Gibbs Fields, Monte Carlo Simulation, and Queues*, Springer-Verlag, 1999.
KELLY F., *Reversibility and Stochastic Networks*, Wiley, 1979.

KLEINROCK L., *Queueing Systems: Volume I – Theory*, Wiley Interscience, 1975.

ROSS K.W., *Multiservice Loss Networks for Broadband Telecommunications Networks*, Springer-Verlag, 1995.

SERFOZO R., *Introduction to Stochastic Networks*, Springer-Verlag, 1999.

Chapter 2

Exponential Distribution

We start with the definition and main properties of the exponential distribution, which is key to the study of Poisson and Markov processes.

2.1. Definition

We say that a non-negative random variable X has the exponential distribution with parameter $\lambda > 0$ if:

$$P(X > t) = e^{-\lambda t}, \quad \forall t \in \mathbb{R}_+.$$

The density of this distribution is given by:

$$f(t) = \lambda e^{-\lambda t}, \quad \forall t \in \mathbb{R}_+.$$

---

[1] *I have an admirable memory, I forget everything.*

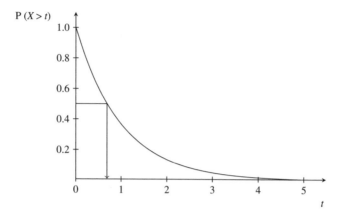

Figure 2.1. *Exponential distribution with parameter* $\lambda = 1$ *and half-life*

The mean and variance of X are, respectively, given by:

$$E(X) = \int_0^\infty t f(t) \, dt = \frac{1}{\lambda},$$

$$\text{var}(X) = \int_0^\infty t^2 f(t) \, dt - E(X)^2 = \frac{1}{\lambda^2}.$$

The exponential distribution is used, for instance, in physics to represent the lifetime of a particle. The parameter λ then corresponds to the rate at which the particle ages. The *half-life* of the particle is defined as the time t such that $P(X > t) = 1/2$, that is $t = \ln(2)/\lambda$, as illustrated in Figure 2.1.

2.2. Discrete analog

The exponential distribution is in continuous time while the geometric distribution is in discrete time. A positive integer random variable X has the geometric distribution with parameter $p \in (0, 1]$ if:

$$P(X = n) = p(1 - p)^{n-1}, \quad \forall n \geq 1,$$

or, equivalently, if:

$$P(X > n) = (1 - p)^n, \quad \forall n \in \mathbb{N}.$$

The mean and variance of X are, respectively, given by:

$$E(X) = \sum_{n=1}^{\infty} np(1 - p)^{n-1} = \frac{1}{p},$$

$$\text{var}(X) = \sum_{n=1}^{\infty} n^2 p(1 - p)^{n-1} - E(X)^2 = \frac{1 - p}{p^2}.$$

Thus, if p represents the probability of winning the lottery, X gives the distribution of the number of attempts necessary to win. When p is low, the geometric distribution is close to the exponential distribution, as illustrated in Figure 2.2.

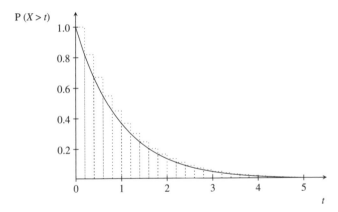

Figure 2.2. *Approximation of the exponential distribution by the geometric distribution*

Formally, denote by $X^{(\tau)}$ a geometric random variable with parameter $p^{(\tau)} = \lambda\tau$, where λ is a fixed, positive parameter and τ a sufficiently small time step. When τ tends to zero, the real random variable $X^{(\tau)}\tau$ tends, in distribution, to an exponential random variable with parameter λ:

$$P(X^{(\tau)}\tau > t) = (1 - p^{(\tau)})^{\lfloor \frac{t}{\tau} \rfloor} \to e^{-\lambda t}, \quad \forall t \in \mathbb{R}_+.$$

2.3. An amnesic distribution

The geometric distribution is *memoryless*: the number of attempts necessary to win the lottery is independent of past attempts. This amnesic property is also satisfied by the exponential distribution:

$$P(X > s + t \mid X > s) = P(X > t), \quad \forall s, t \in \mathbb{R}_+.$$

This is illustrated in Figure 2.3: given the event $X > s$, the random variable $X - s$ has an exponential distribution with parameter λ.

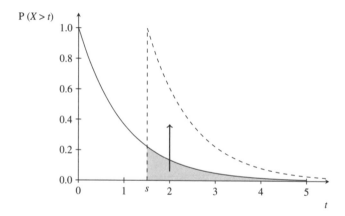

Figure 2.3. *Memoryless distribution: the random variable forgets its past*

Denoting by $F(t) = P(X > t)$ the inverse cumulative distribution function of the random variable X, and observing that for all $s \in \mathbb{R}_+$ such that $F(s) > 0$,

$$P(X > s + t \mid X > s) = \frac{F(s + t)}{F(s)},$$

the amnesic property is equivalent to the functional equation:

$$F(s + t) = F(s)F(t), \quad \forall s, t \in \mathbb{R}_+.$$

The exponential functions are the only solutions to this equation. Since $F(0) = 1$ and F is decreasing, there exists a constant $\lambda > 0$ such that:

$$F(t) = e^{-\lambda t}, \quad \forall t \in \mathbb{R}_+.$$

A consequence of this amnesic property is that an exponentially distributed random variable can be described by its behavior at time $t = 0$. Thus, if X represents the lifetime of a particle, this particle "dies" at constant rate λ, independently of its age:

$$\mathrm{P}(X \leq t) = 1 - e^{-\lambda t} = \lambda t + o(t).$$

2.4. Minimum of exponential variables

Let $X_1, \ldots, X_K$ be K independent exponential random variables respective parameters $\lambda_1, \ldots, \lambda_K$. We denote by λ the sum of these parameters. The minimum X of these random variables satisfies:

$$\mathrm{P}(X > t, X = X_1) = \mathrm{P}(X_1 > t, X_2 \geq X_1, \ldots, X_K \geq X_1),$$

$$= \int_t^\infty \lambda_1 e^{-\lambda_1 s} e^{-\lambda_2 s} \ldots e^{-\lambda_K s} \, ds,$$

$$= \int_t^\infty \lambda_1 e^{-\lambda s} \, ds,$$

$$= \frac{\lambda_1}{\lambda} e^{-\lambda t}, \quad \forall t \in \mathbb{R}_+.$$

Thus the random variable X has an exponential distribution with parameter λ and is equal to X_1 with probability λ_1/λ, independently of the value of X. Equivalently, a set of K exponential timers of respective expiring rates $\lambda_1, \ldots, \lambda_K$ forms an exponential timer of expiring rate λ, each timer being the first to expire with a probability proportional to its rate.

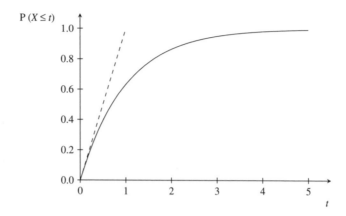

Figure 2.4. *Exponential distribution with parameter λ = 1 and behavior at time t = 0*

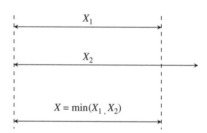

Figure 2.5. *The minimum of independent exponential random variables has an exponential distribution*

This property extends to an infinite sequence of exponential variables $X_1, X_2, \ldots$, provided the sum of their parameters is finite:

$$\lambda = \sum_{k=1}^{\infty} \lambda_k < \infty.$$

The minimum of this sequence has an exponential distribution with parameter λ and is equal to the kth element of the sequence with probability λ_k / λ.

2.5. Sum of exponential variables

Let $X_1, X_2, \ldots,$ be a sequence of i.i.d. exponential random variables with parameter λ. We are interested in the sum S_n of the first n terms of this sequence, with $S_0 = 0$. For any $t > 0$, we denote by $N(t)$ the integer n such that $S_n \le t < S_{n+1}$. This random variable has a Poisson distribution with mean λt:

$$
\begin{aligned}
\mathrm{P}(N(t) = n) \\
&= \mathrm{P}(S_n \le t < S_{n+1}), \\
&= \int_{t_1+\ldots+t_n \le t} \lambda^n e^{-\lambda t_1} \ldots e^{-\lambda t_n} e^{-\lambda(t-t_1-\ldots-t_n)} \, dt_1 \ldots dt_n, \\
&= \lambda^n e^{-\lambda t} \int_{t_1+\ldots+t_n \le t} dt_1 \ldots dt_n, \\
&= \frac{(\lambda t)^n}{n!} e^{-\lambda t}.
\end{aligned}
$$

We deduce:

$$
\mathrm{P}(S_n > t) = \sum_{k=0}^{n-1} \mathrm{P}(N(t) = k) = e^{-\lambda t}\left(1 + \lambda t + \ldots + \frac{(\lambda t)^{n-1}}{(n-1)!}\right).
$$

Thus, a series of n successive exponential timers of expiring rate λ has an Erlang distribution[2] with parameters n, λ.

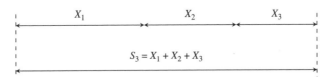

Figure 2.6. *The sum of exponential random variables has an Erlang distribution*

---

2 The Erlang distribution is a Gamma distribution whose form parameter, here n, is an integer.

2.6. Random sum of exponential variables

Now consider the sum S of the first N terms of the sequence $X_1, X_2, \ldots$, where N is a geometrically distributed random variable with parameter $p \in (0, 1]$. Thus, the random variable S is equal to X_1 with probability p, to $X_1 + X_2$ with probability $p(1 - p)$, etc. The random variable S has an exponential distribution with parameter $p\lambda$:

$$\mathrm{P}(S > t) = \sum_{n=1}^{\infty} p(1-p)^{n-1} \mathrm{P}(S_n > t),$$

$$= \sum_{n=1}^{\infty} p(1-p)^{n-1} \sum_{k=0}^{n-1} e^{-\lambda t} \frac{(\lambda t)^k}{k!},$$

$$= \sum_{k=0}^{\infty} e^{-\lambda t} \frac{(\lambda t)^k}{k!} \sum_{n>k} p(1-p)^{n-1},$$

$$= \sum_{k=0}^{\infty} e^{-\lambda t} \frac{((1-p)\lambda t)^k}{k!} = e^{-p\lambda t}, \quad \forall t \in \mathbb{R}_+.$$

Thus, an exponential timer of rate λ that is reinitiated with probability $1 - p$ at each expiring time forms an exponential timer of rate $p\lambda$.

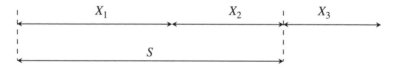

Figure 2.7. *The sum of a geometric number of exponential random variables has an exponential distribution*

2.7. A limiting distribution

Let X be a positive real random variable whose inverse cumulative distribution function $F(t) = \mathrm{P}(X > t)$ satisfies

$F(0) = 1$ and has a negative, finite derivative in $t = 0$. For all $n \geq 1$, let X_n be the minimum of n i.i.d. random variables with the same distribution as nX. Then X_n tends, in distribution, to an exponential random variable with parameter $\lambda = -F'(0)$ when n tends to infinity:

$$P(X_n > t) = P(nX > t)^n = F\left(\frac{t}{n}\right)^n \to e^{-\lambda t}, \quad \forall t \in \mathbb{R}_+.$$

Thus, a large number of i.i.d. "slow" random timers form an exponential timer whose rate depends on the expiring rate of these timers at time $t = 0$.

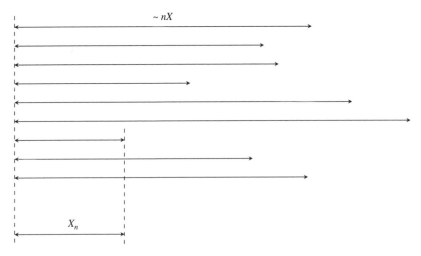

Figure 2.8. *The minimum of a large number of "slow" random variables has an exponential distribution*

2.8. A "very" random variable

The exponential distribution is the distribution of maximal entropy among all distributions with the same mean having a density on $\mathbb{R}_+$. The entropy of a positive real random variable X with density f is defined by:

$$h(X) = -\int_0^\infty f(t) \ln(f(t)) \, \mathrm{d}t.$$

Thus, the entropy of an exponential random variable X with parameter $\lambda > 0$ is given by:

$$h(X) = -\ln(\lambda) + \int_0^\infty \lambda t \times \lambda e^{-\lambda t}\, dt = 1 - \ln(\lambda).$$

Now for all density functions f, g on $\mathbb{R}_+$, it follows from Gibbs inequality that:

$$\int_0^\infty f(t) \ln\left(\frac{f(t)}{g(t)}\right) dt \geq 0,$$

with equality if and only if $f = g$. Applying this inequality to $g(t) = \lambda e^{-\lambda t}$, we obtain for any random variable X with density f and mean $1/\lambda$:

$$h(X) = -\int_0^\infty f(t) \ln(f(t))\, dt \leq -\int_0^\infty f(t) \ln(g(t))\, dt = 1 - \ln(\lambda),$$

with equality if and only if X is an exponential random variable.

2.9. Exercises

1. *Life of a particle*
    A particle has an exponential lifetime with parameter λ. Calculate its half-life. What is the expected lifetime of the particle after its half-life?

2. *Discretization*
    Let X be an exponential random variable with parameter λ. Show that $\lceil \frac{X}{\tau} \rceil$ has a geometric distribution with parameter $1 - e^{-\lambda \tau}$, for any $\tau > 0$. Give the expectation of $\tau \lceil \frac{X}{\tau} \rceil$ when τ is equal to the half-life of X.

3. Lighting

Two bulbs have average lifetimes of 1,000 and 2,000 hours. Assuming that the lifetime of each bulb has an exponential distribution, when is it necessary to replace one of the two bulbs on average? What is the probability that the "long-life" bulb must be replaced first?

4. Law of the minimum

Let X_1 and X_2 be two independent geometric random variables with respective parameters p_1 and p_2. What is the distribution of the random variable $X = \min(X_1, X_2)$? Use this result to derive the distribution of the minimum of exponential random variables.

5. Rare events

What is the approximate distribution of the minimum of 100 independent random variables uniformly distributed over $[0, 100]$?

6. Synchronization

Let X_1 and X_2 be two independent exponential random variables with respective parameters λ_1 and λ_2. What is the probability that $X_1 = X_2$?

7. Waiting time of a bus

The waiting time of a bus X has a Pareto distribution with parameter $\alpha > 0$:

$$P(X > t) = \left(\frac{\beta}{t}\right)^\alpha, \ \forall t > \beta$$

for some $\beta > 0$. Show that the probability that $X > s + t$, given $X > s$, is an increasing function of s for all $t > 0$. In other words, the longer you wait for the bus, the longer you are likely to wait further. Is this still the case if X has a uniform distribution on $[0, 1]$?

2.10. Solution to the exercises

1. Life of a particle

Let X be the lifetime of the particle. Since X has an exponential distribution with parameter λ, its half-life is given by $t = \ln(2)/\lambda$. The exponential distribution is memoryless (see section 2.3), so the expected lifetime of the particle after its half-life is unchanged, equal to $1/\lambda$.

2. Discretization

We have:

$$\forall n \geq 1, \quad P\left(\left[\frac{X}{\tau}\right] = n\right) = P(n - 1 < X/\tau \leq n),$$

$$= \int_{(n-1)\tau}^{n\tau} \lambda e^{-\lambda t}\, dt,$$

$$= (1 - e^{-\lambda\tau})e^{-(n-1)\lambda\tau}.$$

This is a geometric distribution with parameter $1 - e^{-\lambda\tau}$. For $\tau = \ln(2)/\lambda$, we get:

$$E\left(\tau\left[\frac{X}{\tau}\right]\right) = \frac{2\ln(2)}{\lambda},$$

There is a bias because of the relatively large value of τ. This bias disappears when $\tau \to 0$.

3. Lighting

We use here the property of the minimum of two exponential variables presented in section 2.4. The lifetimes of the bulbs having exponential distributions with respective parameters $\lambda_1 = 1/1{,}000\,\mathrm{h}^{-1}$ and $\lambda_2 = 1/2{,}000\,\mathrm{h}^{-1}$, one of the two bulbs must be replaced after an exponential time with parameter $\lambda = \lambda_1 + \lambda_2 = 3/2{,}000\,\mathrm{h}^{-1}$, that is, $2{,}000/3 \approx 666$ hours on average. The probability that the long-life bulb must be replaced first is equal to λ_2/λ, that is, $1/3$.

4. Law of the minimum

Using the independence of X_1 and X_2, we obtain:

$$P(X > n) = P(X_1 > n, X_2 > n),$$
$$= P(X_1 > n)P(X_2 > n),$$
$$= ((1 - p_1)(1 - p_2))^n, \ \forall n \in \mathbb{N}.$$

So X has a geometric distribution with parameter $p_1 + p_2 - p_1 p_2$.

As given in section 2.2, we define two random variables $X_1^{(\tau)}$ and $X_2^{(\tau)}$ of geometric distributions with respective parameters $\lambda_1 \tau$ and $\lambda_2 \tau$, for any sufficiently small τ. We know that the distributions of these random variables tend to exponential distributions with respective parameters λ_1 and λ_2 when τ tends to 0. We then define $X^{(\tau)} = \min(X_1^{(\tau)}, X_2^{(\tau)})$. We have for all $t > 0$:

$$P(X^{(\tau)}\tau > t) = ((1-\lambda_1\tau)(1-\lambda_2\tau))^{\lfloor \frac{t}{\tau} \rfloor} \to e^{-(\lambda_1+\lambda_2)t} \quad \text{when } \tau \to 0.$$

Thus, the distribution of $X^{(\tau)}\tau$ tends to an exponential distribution with parameter $\lambda = \lambda_1 + \lambda_2$.

5. Rare events

In view of section 2.7, the minimum of 100 independent random variables uniformly distributed over $[0, 100]$ is approximately distributed as an exponential random variable with parameter 1.

6. Synchronization

We have:

$$P(X_1 = X_2) = \int \int 1(t_1 = t_2)\lambda_1 e^{-\lambda_1 t_1}\lambda_2 e^{-\lambda_2 t_2} \, dt_1 \, dt_2,$$
$$\leq \lambda_1\lambda_2 \int \int 1(t_1 = t_2) \, dt_1 \, dt_2,$$
$$= 0.$$

Thus, X_1 and X_2 are a.s. distinct. This property holds for any independent real-valued random variables X_1, X_2 having densities.

7. *Waiting time of a bus*

It is sufficient to observe that:

$$P(X > s + t \mid X > s) = \left(\frac{s}{s+t} \right)^{\alpha},$$

which increases with s.

For a uniform distribution on $[0, 1]$, we have $P(X > t) = 1 - t$ for all $t \in [0, 1]$ so that:

$$P(X > s + t \mid X > s) = \frac{1 - t - s}{1 - s}.$$

This probability decreases with s.

Chapter 3

Poisson Processes

We now introduce the Poisson[1] process, describing the times of random events, such as the arrivals of customers in a queue. We shall see that this process is simply produced by a large number of rare, mutually independent events. This explains why Poisson processes are so common in physical, biological, and computer systems.

3.1. Definition

A *point process* on $\mathbb{R}_+$ is an increasing sequence $T_1, T_2, T_3, \ldots$, of non-negative, real random variables representing the times of arbitrary events. For convenience, we shall refer to these events as *arrivals*. Letting $T_0 = 0$, we denote by $\tau_1, \tau_2, \tau_3, \ldots$, with $\tau_n = T_n - T_{n-1}$ the *inter-arrival times* of the process. A point process is said to be *simple* if its inter-arrival times are a.s. positive and *stationary* if for all $n \geq 0$, the sequence of inter-arrival times $\tau_{n+1}, \tau_{n+2}, \tau_{n+3}, \ldots$, has the same distribution as the initial sequence $\tau_1, \tau_2, \tau_3, \ldots$.

---

1 Siméon Denis Poisson, French mathematician and physicist (1781–1840).

We then define the *intensity* of the process as the quantity:

$$\lambda = \frac{1}{\mathrm{E}(\tau_1)}.$$

A Poisson process of intensity $\lambda > 0$ is a point process whose inter-arrival times form a sequence of independent *exponential* random variables with parameter λ. It is a simple and stationary point process.

Any point process can also be described by its counting measure N, defined by:

$$N(t) = \sum_{n \geq 1} 1(T_n \leq t), \quad \forall t \in \mathbb{R}_+.$$

For all s, t such that $s < t$, we denote by $N(s, t) = N(t) - N(s)$ the number of arrivals in the interval $(s, t]$. The process is simple if the probability that $N(s, t) > 1$ given that $N(s, t) \geq 1$ tends to 0 when s tends to t and stationary if the random variable $N(s, t)$ has the same distribution as $N(0, t - s)$ for all s, t such that $s < t$. The intensity of the process is then given by the quantity:

$$\lambda = \mathrm{E}(N(0, 1)).$$

A Poisson process of intensity $\lambda > 0$ is a point process such that the numbers of arrivals in any disjoint intervals are independent and the number of arrivals in any interval $(s, t]$ has a *Poisson distribution*[2] with mean $\lambda(t - s)$:

$$P(N(s, t) = n) = e^{-\lambda(t-s)} \frac{(\lambda(t - s))^n}{n!}, \quad \forall n \in \mathbb{N}.$$

---

2 This is the "law of small numbers", as described by the Russian mathematician Ladislaus Josephowitsch Bortkiewicz (1868–1931) in *The Law of Small Numbers*, 1898. He verified this law on the number of soldiers of the Prussian army killed every year by horse fall.

Both definitions are equivalent, as shown below in the discrete version of Poisson processes.

Note that the intensity λ corresponds to the *arrival rate* of the Poisson process. The numbers of arrivals in the intervals $(0,1], (1,2], \ldots$, being i.i.d. random variables with mean λ, we indeed have by the strong law of large numbers:

$$\lim_{t \to \infty} \frac{1}{t} N(t) = E(N(0,1)) = \lambda \quad \text{a.s.}$$

3.2. Discrete analog

A point process on $\mathbb{N}$ is an increasing sequence $S_1, S_2, S_3, \ldots$, of non-negative integer random variables. Letting $S_0 = 0$, we denote by $\sigma_1, \sigma_2, \sigma_3, \ldots$, with $\sigma_n = S_n - S_{n-1}$ the inter-arrival times of this process. Again, the point process is said to be *simple* if its inter-arrival times are a.s. positive and *stationary* if for all $n \geq 0$, the sequence of inter-arrival times $\sigma_{n+1}, \sigma_{n+2}, \sigma_{n+3}, \ldots$, has the same distribution as the initial sequence $\sigma_1, \sigma_2, \sigma_3, \ldots$. We then define the *intensity* of the process as the quantity:

$$p = \frac{1}{E(\sigma_1)}.$$

A Bernoulli[3] sequence of intensity $p \in (0,1]$ is a point process on $\mathbb{N}$ whose inter-arrival times form a sequence of independent *geometric* random variables with parameter p. It is a simple and stationary point process.

The point process can also be defined by counting the arrivals. For all $k \geq 1$, we denote by X_k the number of arrivals at time k:

$$X_k = \sum_{n \geq 1} 1(S_n = k).$$

---

3 Jacques Bernoulli, Swiss mathematician and physicist (1654–1705).

The process is simple if $X_k \leq 1$ a.s. and stationary if X_k has the same distribution as X_1 for all k. The intensity of the process is then given by the quantity:

$$p = E(X_1).$$

A Bernoulli sequence of intensity $p \in (0, 1]$ is a point process on $\mathbb{N}$ such that the random variables $X_1, X_2, \ldots$, are independent and have a *Bernoulli* distribution with mean p:

$$P(X_k = 1) = p, \quad P(X_k = 0) = 1 - p, \quad \forall k \geq 1.$$

Both definitions are equivalent.

When p is low, the Bernoulli sequence is stochastically close to a Poisson process, as illustrated in Figure 3.1.

Figure 3.1. *The Bernoulli sequence is the discrete-time version of a Poisson process*

Formally, we denote by $S_1^{(\tau)}, S_2^{(\tau)}, \ldots$, a Bernoulli sequence of intensity $p^{(\tau)} = \lambda\tau$, where λ is a fixed, positive parameter and τ some sufficiently small time step. When τ tends to zero, the inter-arrival times are independent and tend, in distribution, to exponential variables with parameter λ (see section 2.2); the limiting process is thus a Poisson process of intensity λ. We obtain the same result by observing that the number of arrivals in any disjoint intervals is independent and that, in the limit, the number of arrivals in any interval $(s, t]$ has a Poisson distribution with mean $\lambda(t - s)$:

$$P(N^{(\tau)}(s, t) = n) = \binom{\lfloor \frac{t}{\tau} \rfloor - \lfloor \frac{s}{\tau} \rfloor}{n} p^{(\tau)n} (1 - p^{(\tau)})^{\lfloor \frac{t}{\tau} \rfloor - \lfloor \frac{s}{\tau} \rfloor - n},$$

$$\rightarrow \frac{(\lambda(t - s))^n}{n!} e^{-\lambda(t-s)}.$$

3.3. An amnesic process

A Poisson process is memoryless: the knowledge of the process until time t does not say anything about its future evolution. This property characterizes the Poisson processes among all simple and stationary point processes.

This can be easily verified in discrete time. Consider a memoryless, simple, and stationary point process. The fact that there is an arrival at time k is independent of the arrivals before time k; since the process is stationary, there is an arrival at time k with some probability $p > 0$, which is independent of k: the point process is a Bernoulli sequence of intensity p.

3.4. Distribution of the points of a Poisson process

Knowing that some interval $(s, t]$ of $\mathbb{R}_+$ contains n points of a Poisson process, these points are uniformly distributed over this interval. We have indeed, for any positive integer l, any partition of $(s, t]$ in l intervals $(s, t_1], (t_1, t_2], \ldots, (t_{l-1}, t]$ and any integers $n_1, \ldots, n_l$ of sum equal to n:

$$P(N(s, t_1)$$
$$= n_1, \ldots, N(t_{l-1}, t) = n_l \mid N(s, t) = n)$$
$$= \frac{P(N(s, t_1) = n_1, \ldots, N(t_{l-1}, t) = n_l)}{P(N(s, t) = n)},$$
$$= \frac{e^{-\lambda(t_1 - s)} \dfrac{(\lambda(t_1 - s))^{n_1}}{n_1!} \ldots e^{-\lambda(t - t_{l-1})} \dfrac{(\lambda(t - t_{l-1}))^{n_l}}{n_l!}}{e^{-\lambda(t - s)} \dfrac{(\lambda(t - s))^{n}}{n!}},$$
$$= \binom{n}{n_1, \ldots, n_l} \left(\frac{t_1 - s}{t - s}\right)^{n_1} \ldots \left(\frac{t - t_{l-1}}{t - s}\right)^{n_l}.$$

The probability that any of these l intervals contains one of the n points of the Poisson process is thus proportional to its length. We deduce that each point is uniformly distributed on the interval $(s, t]$, independently of the other points.

3.5. Superposition of Poisson processes

The superposition of K independent Poisson processes of respective intensities $\lambda_1, \dots, \lambda_K$ is a Poisson process of intensity $\lambda = \lambda_1 + \dots + \lambda_K$. Denoting by $N_1, \dots, N_K$ the counting measures of these K Poisson processes, the counting measure N associated with the superposed process indeed verifies:

$$P(N(s, t) = n)$$

$$= \sum_{\substack{n_1, \dots, n_K \\ n_1 + \dots + n_K = n}} P(N_1(s, t) = n_1) \dots P(N_K(s, t) = n_K),$$

$$= \sum_{\substack{n_1, \dots, n_K \\ n_1 + \dots + n_K = n}} \frac{(\lambda_1(t - s))^{n_1}}{n_1!} e^{-\lambda_1(t-s)} \dots \frac{(\lambda_K(t - s))^{n_K}}{n_K!} e^{-\lambda_K(t-s)},$$

$$= \sum_{\substack{n_1, \dots, n_K \\ n_1 + \dots + n_K = n}} \binom{n}{n_1, \dots, n_K} (\lambda_1(t - s))^{n_1} \dots (\lambda_K(t - s))^{n_K} e^{-\lambda(t-s)},$$

$$= \frac{\lambda^n(t - s)^n}{n!} e^{-\lambda(t-s)},$$

and the property of independence of the counting measures over disjoint intervals follows from that satisfied by the original Poisson processes.

REMARK 3.1.– Since a Poisson process is simple, it follows from the above property that two independent Poisson processes have a.s. no common points.

Moreover, the probability that some arbitrary point of the superposed process belongs to one of the original K processes is proportional to the intensity of this process:

$$P(N_1(s,t) = 1 \mid N(s,t) = 1)$$

$$= \frac{P(N_1(s,t) = 1, N_2(s,t) = 0, \ldots, N_K(s,t) = 0)}{P(N(s,t) = 1)},$$

$$= \frac{\lambda_1(t-s)e^{-\lambda(t-s)}}{\lambda(t-s)e^{-\lambda(t-s)}} = \frac{\lambda_1}{\lambda}.$$

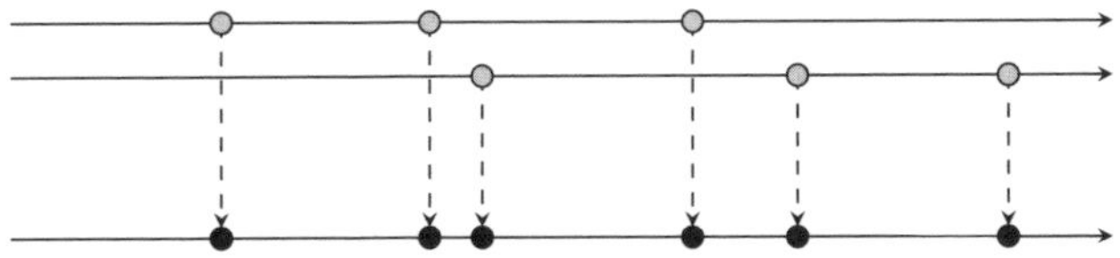

Figure 3.2. *Superposition of Poisson processes*

3.6. Subdivision of a Poisson process

Let N be the counting measure of a Poisson process of intensity λ and $p_1, \ldots, p_K$ be K positive real numbers of sum equal to 1. The K point processes obtained by distributing each point of the Poisson process to the kth point process with probability p_k are independent Poisson processes of respective intensities $\lambda p_1, \ldots, \lambda p_K$. The corresponding counting measures $N_1, \ldots, N_K$ indeed verify for any interval $(s, t]$:

$$P(N_1(s,t) = n_1, \ldots, N_K(s,t) = n_K)$$

$$= \binom{n}{n_1, \ldots, n_K} p_1^{n_1} \ldots p_K^{n_K} \frac{\lambda^n(t-s)^n}{n!} e^{-\lambda(t-s)},$$

$$= \frac{(\lambda p_1(t-s))^{n_1}}{n_1!} e^{-\lambda p_1(t-s)} \ldots \frac{(\lambda p_K(t-s))^{n_K}}{n_K!} e^{-\lambda p_K(t-s)},$$

with $n = n_1 + \ldots + n_K$, and the property of independence of the counting measures over disjoint intervals follows from that satisfied by the original Poisson process.

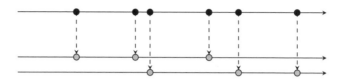

Figure 3.3. *Subdivision of a Poisson process*

3.7. A limiting process

For any simple, stationary point process of intensity $\lambda > 0$, the point process obtained by superposing n i.i.d. versions of this process, each dilated by a factor n, tends, in distribution, to a Poisson process of intensity λ when n tends to infinity. We explain this property through the simple case of periodic processes.

Consider the superposition of n periodic processes of intensity λ/n with independent random phases, uniformly distributed over $[0, n/\lambda]$. For large n, each periodic process has at the most one point in any fixed set of disjoint intervals; the property of independence of the counting measure of the superposed process over disjoint intervals follows. Moreover, each periodic process has a point in some time interval of duration t with probability $(\lambda t)/n$, independently of the other processes. Thus, the number of points of the superposed process in this interval has a binomial distribution with parameters $n, (\lambda t)/n$. This tends to a Poisson distribution with mean λt for large n. Thus, the limiting process is a Poisson process of intensity λ.

3.8. A "very" random process

Finally, the Poisson process is the simple stationary point process of maximum entropy. We show this property in discrete time, by looking for all $n \geq 1$ at the joint entropy of the random variables $X_1, \ldots, X_n$, with $X_k = 1$, if there is an arrival at time k, $X_k = 0$ otherwise. Letting $p = P(X_k = 1)$, which is independent of k by stationarity, we obtain:

$$H(X_1, \ldots, X_n) \leq \sum_{k=1}^{n} H(X_k) = -n(p \ln(p) + (1-p) \ln(p)),$$

with equality when the random variables $X_1, \ldots, X_n$ are independent, that is, for a Bernoulli sequence.

3.9. Exercises

1. Counting

Let N be the counting measure of a Poisson process of intensity λ. Calculate the mean and the variance of the random variable $N(t)$.

2. Bank

Consider customers arriving at a bank according to a Poisson process. There is one arrival every 10 minutes on average. What is the probability that no customer arrives during one hour? Given that there have been no arrivals during the last half-hour, what is the probability that there will be no arrivals during the next half-hour? What is the probability of having at least two arrivals in less than one minute?

3. Toll

A toll consists of 10 lanes. The vehicles arrive according to a Poisson process and are randomly, uniformly distributed over these lanes. If the traffic intensity is equal to 1,000 vehicles per hour, what is the probability that the inter-arrival time between two vehicles at any given lane is larger than one minute?

4. *Museum*

It takes one hour to visit a museum. Visitors arrive according to a Poisson process of intensity λ. Assuming that the museum is open for more than one hour, give the distribution of the number of visitors in the museum.

5. *The bus paradox*

Buses arrive according to a Poisson process of intensity λ at some stop. What is the average waiting time of a bus? How much time has elapsed since the last bus? Deduce that, from the viewpoint of a user arriving at some arbitrary time, the mean bus inter-arrival time is equal to double the actual mean bus inter-arrival time.

More generally, show that if buses arrive every X time units, where X is a random variable with density f and finite mean, the average time between two buses is equal to $E(X^2)/E(X)$ from the viewpoint of a user; we will admit that, from the viewpoint of a user, the time between two buses has the density $x \mapsto xf(x)$, up to some multiplicative constant.

6. *Interleaving*

Given two Poisson processes of respective intensities λ_1 and λ_2, calculate the average number of arrivals of one of these processes between two arrivals of the other.

3.10. Solution to the exercises

1. *Counting*

The random variable $N(t)$ has a Poisson distribution with parameter λt. Both its mean and variance are given by λt.

2. *Bank*

The number of arrivals in any interval of duration t has a Poisson distribution with parameter λt, with $\lambda = 1/10 \, \mathrm{min}^{-1}$.

The probability that there are no arrivals in $t = 60\,\text{min}$ is equal to:

$$P(N(t) = 0) = e^{-\lambda t} = e^{-6} \approx 2.5 \times 10^{-3}.$$

The arrivals in two disjoint intervals being independent, the probability that there are no arrivals during $t = 30\,\text{min}$ is independent of the arrivals in the previous half-hour and is given by:

$$P(N(t) = 0) = e^{-\lambda t} = e^{-3} \approx 5.0 \times 10^{-2}.$$

The probability of having at least two arrivals during $t = 1\,\text{min}$ is given by:

$$P(N(t) \geq 2) = 1 - P(N(t) \leq 1) = 1 - e^{-\lambda t}(1 + \lambda t),$$

$$= 1 - e^{-1/10}(1 + 1/10) \approx 4.7 \times 10^{-3}.$$

3. Toll

Each vehicle chooses its lane uniformly at random, independently of the other vehicles. Applying the principle of subdivision of Poisson processes presented in section 3.6, we obtain that the process of vehicle arrivals at each lane is a Poisson process of intensity $\lambda = 100$ vehicles per hour. The probability that the inter-arrival time between two vehicles is larger than one minute is equal to the probability that there are no arrivals during $t = 1\,\text{min}$, that is:

$$P(N(t) = 0) = e^{-\lambda t} = e^{-5/3} \approx 0.19.$$

4. Museum

The number of visitors in the museum at time t corresponds to the number of arrivals of the Poisson process in the interval $(t - 1, t]$ (the visitors who arrived before time $t - 1$ have left the museum). We obtain a Poisson distribution with parameter λ.

5. *The bus paradox*

From the memoryless property of Poisson processes, the waiting time of bus has an exponential distribution with parameter λ. By symmetry, the elapsed time since the arrival of the previous bus also has an exponential distribution with parameter λ. From the viewpoint of the user, the mean bus inter-arrival time is thus equal to $2/\lambda$, instead of $1/\lambda$. This paradox can be explained by the fact that a user arriving at an arbitrary time at the bus stop is more likely to arrive during a long interval between two buses.

More generally, the mean bus inter-arrival time has, from the viewpoint of the user, density $x \mapsto xf(x)/\mathrm{E}(X)$. The corresponding mean is equal to $\mathrm{E}(X^2)/\mathrm{E}(X)$, which is larger than the actual mean bus inter-arrival time, $\mathrm{E}(X)$.

6. *Interleaving*

We use the property of superposition of the Poisson processes, see section 3.5. The superposition of the two processes forms a Poisson process of intensity $\lambda = \lambda_1 + \lambda_2$, each point of which belongs to the first process with probability λ_1/λ and to the second process with probability λ_2/λ. The number X of arrivals of the first process between two arrivals of the second thus has a geometric distribution with parameter λ_2/λ on $\mathbb{N}$:

$$\forall n \geq 0, \quad \mathrm{P}(X = n) = \frac{\lambda_2}{\lambda} \left(\frac{\lambda_1}{\lambda} \right)^n.$$

Chapter 4

Markov Chains

While the probability theory has long been restricted to the analysis of sequences of independent events (the successive flips of a coin for instance), Markov[2] has significantly enlarged its application field by stating the main properties of sequences of *correlated* events. The key component of his theory is known as the *Markov property*: given its present state (at time n), the future state of the sequence (at time $n + 1$) must be independent of its past states (at times $n - 1$ and before). We start with Markov chains, in discrete time; the case of Markov *processes*, in continuous time, is considered in Chapter 5.

---

1 *The future is the past in preparation.*
2 Andreï Andreïevich Markov, Russian mathematician (1856–1922).

4.1. Definition

We say that a sequence of random variables $X_0, X_1, \ldots,$ with values in some countable space $\mathcal{X}$ is a Markov chain if this sequence is memoryless in the sense that for all $n \geq 0$ and all states $x, y \in \mathcal{X}, x_0, x_1, \ldots, x_{n-1} \in \mathcal{X}$:

$$P(X_{n+1} = y \mid X_n = x; X_0 = x_0, \ldots, X_{n-1} = x_{n-1})$$

$$= P(X_{n+1} = y \mid X_n = x).$$

The Markov chain is said to be *homogeneous* if this expression is independent of n.

4.2. Transition probabilities

In the following, we consider homogeneous Markov chains only. Such a Markov chain is determined by its transition probabilities, defined by:

$$\forall x, y \in \mathcal{X}, \quad p(x, y) = P(X_1 = y \mid X_0 = x).$$

Note that we have:

$$\forall x \in \mathcal{X}, \quad \sum_{y \in \mathcal{X}} p(x, y) = 1. \qquad [4.1]$$

We refer to the *transition graph* as the oriented graph whose nodes are the elements $\mathcal{X}$ and the edges represent the transitions of positive probability, as illustrated in Figure 4.1. The weight of each edge is equal to the probability of the corresponding transition. The Markov chain is said to be *irreducible* if its transition graph is strongly connected, that is, if for all distinct states $x, y \in \mathcal{X}$, there exist paths from x to y and from y to x in the graph. In the following, we assume that the Markov chain is irreducible.

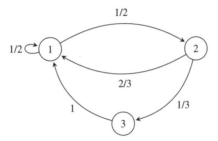

Figure 4.1. *Transition graph of a Markov chain on X = {1,2,3}*

4.3. Periodicity

The period of a Markov chain is the largest common divisor of lengths of cycles in the transition graph. The chain is said to be *aperiodic* if its period is equal to 1.

For instance, the Markov chain on the left side of Figure 4.2 has period 2; if $X_0 = 1$, then $X_n = 1$ when n is even and $X_n \in \{2, 3\}$ when n is odd. The chain has a periodic behavior. On the other hand, the Markov chain on the right side of Figure 4.2 is aperiodic (there are cycles of lengths 2 and 3); it may easily be verified that for any $n \geq 5$, there is a positive probability that $X_n = x$, for all $x \in \{1, 2, 3\}$, whatever the initial state X_0. More generally, an aperiodic Markov chain is such that, for sufficiently large n, the probability that $X_n = x$ is positive for all $x \in \mathcal{X}$, whatever the initial state X_0.

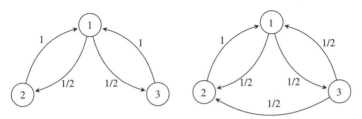

Figure 4.2. *Markov chains of period 2 (left) and 1 (right) on X = {1,2,3}*

REMARK 4.1.– (Loops). A Markov chain that has a loop transition (i.e. some state x such that $p(x, x) > 0$) is aperiodic (there is a cycle of length 1).

4.4. Balance equations

We say that a probability measure π on $\mathcal{X}$ is a *stationary distribution* of the Markov chain if the distribution of X_n is equal to π for all $n \geq 0$ whenever the distribution of the initial state X_0 is equal to π. We then say that the Markov chain is in *steady state*. Writing:

$$\forall x \in \mathcal{X}, \quad \mathrm{P}(X_1 = x) = \sum_{y \in \mathcal{X}} \mathrm{P}(X_0 = y)\mathrm{P}(X_1 = x \mid X_0 = y),$$

we see that the stationary distribution, if it exists, satisfies the equations:

$$\forall x \in \mathcal{X}, \quad \pi(x) = \sum_{y \in \mathcal{X}} \pi(y)p(y, x). \tag{4.2}$$

These are the *balance equations* of the Markov chain.

4.5. Stationary measure

We call *stationary measure* any non-null solution to the balance equations; a stationary measure is defined up to a multiplicative constant and, in particular, is generally not a probability measure.

Note that any stationary measure satisfies:

$$\forall x \in \mathcal{X}, \quad \pi(x) > 0. \tag{4.3}$$

This property follows from irreducibility: since the stationary measure is non-null, there exists some state $x \in \mathcal{X}$ such that $\pi(x) > 0$; it then follows from equation [4.2] that $\pi(y) > 0$ for all states $y \in \mathcal{X}$ such that $p(y, x) > 0$, and, step by step, that $\pi(y) > 0$ for all states $y \in \mathcal{X}$.

There exists a *stationary distribution* if and only if there exists a stationary measure of finite sum. The stationary distribution π is then unique and is obtained from the normalization of the stationary measure, so that:

$$\sum_{x \in \mathcal{X}} \pi(x) = 1. \tag{4.4}$$

4.6. Stability and ergodicity

A Markov chain having a stationary distribution π is said to be *stable*. When the Markov chain is aperiodic, this stationary distribution is also the limiting distribution of the Markov chain in the sense that, for all states x:

$$\lim_{n \to \infty} P(X_n = x) = \pi(x),$$

whatever the initial state X_0. Moreover, the stationary distribution gives the frequency of each state, that is, for all states x:

$$\lim_{N \to \infty} \frac{1}{N} \sum_{n=1}^{N} 1(X_n = x) = \pi(x) \quad \text{a.s.} \tag{4.5}$$

whatever the initial state X_0. This is illustrated in Figure 4.3 for some sample path of the Markov chain of Figure 4.1.

More generally, we have for any function $f \colon \mathcal{X} \to \mathbb{R}$ such that $E(f(X)) < \infty$:

$$\lim_{N \to \infty} \frac{1}{N} \sum_{n=1}^{N} f(X_n) = E(f(X)) \quad \text{a.s.} \tag{4.6}$$

where X denotes some random variable of distribution π. This is the *ergodic theorem*, which states that the means in time (the left-hand side of the equality) and space (the right-hand side of the equality) coincide. This is an extension of the strong law of large numbers to correlated random variables. We say that the Markov chain is *ergodic*.

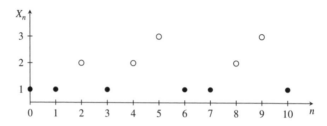

Figure 4.3. *Ergodicity: $\pi(1)$ is the fraction of time spent in state 1*

A Markov chain having no stationary distribution is said to be *unstable*. It then goes to infinity in the sense that, for all states x:

$$\lim_{n \to \infty} P(X_n = x) = 0,$$

whatever the initial state X_0. Thus, the chain eventually leaves any finite subset of states, which implies that the state space is infinite. A Markov chain is unstable if there is no positive solution to the balance equations [4.2] or if the only positive solutions have an infinite sum.

4.7. Finite state space

When the state space $\mathcal{X}$ is finite, the balance equations [4.2] can be written in vectorial notation:

$$\pi = \pi P,$$

where π is a line vector with components $\pi(x)$ and P a stochastic matrix with entries $p(x, y)$. In particular, any stationary measure corresponds to a left eigenvector of the transition matrix P for the eigenvalue 1. The existence and uniqueness (up to some multiplicative constant) of the stationary measure then follows from Perron–Frobenius' theorem[3].

The Markov chain is always stable. Specifically, if the state space has cardinal N, any set of $N - 1$ equations among the N balance equations [4.2] forms, with the normalization condition [4.4], a linear system of N independent equations with N unknown variables (see exercise 1 in section 4.16).

4.8. Recurrence and transience

The standard classification of Markov chains is based on the notions of recurrence and transience: a chain is said to be *recurrent* if the return time to any state is a.s. finite, and *transient* otherwise. Moreover, a recurrent chain is said to be *positive* if the return time is of finite mean, and *null* otherwise. According to our terminology, a stable chain is positive recurrent, whereas an unstable chain is either null recurrent or transient. For a stable chain, the mean return time to any state x (starting from this state) is the inverse of the frequency of state x that is in view of the ergodic theorem [4.5]:

$$\frac{1}{\pi(x)}. \qquad [4.7]$$

---

[3] Oskar Perron (1880–1975) and Ferdinand Georg Frobenius (1849–1917), German mathematicians.

4.9. Frequency of transition

For any pair of states $x, y \in \mathcal{X}$, the quantity $\pi(x)p(x,y)$ corresponds to the *frequency of transition* from state x to state y, that is:

$$\pi(x)p(x,y) = \lim_{N \to \infty} \frac{1}{N} \sum_{n=1}^{N} 1(X_n = x, X_{n+1} = y) \quad \text{a.s.} \quad [4.8]$$

This result is obtained from the application of the ergodic theorem [4.5] to the sequence (X_0, X_1), $(X_1, X_2), \ldots$ This sequence indeed forms a Markov chain whose stationary distribution is obtained by considering the original Markov chain in steady state:

$$P(X_0 = x, X_1 = y) = P(X_0 = x)P(X_1 = y \mid X_0 = x),$$
$$= \pi(x)p(x,y), \quad \forall x, y \in \mathcal{X}.$$

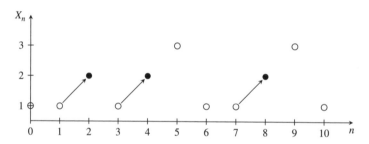

Figure 4.4. *Frequency of transition from state 1 to state 2, given by π(1)p(1,2)*

In particular, the arrival frequency to state x and the departure frequency from state x are, respectively, given by:

$$\sum_{y \in \mathcal{X}} \pi(y)p(y,x) \quad \text{and} \quad \sum_{y \in \mathcal{X}} \pi(x)p(x,y).$$

These quantities, which correspond to the frequency of visits of state x, are equal:

$$\sum_{y \in \mathcal{X}} \pi(y)p(y,x) = \sum_{y \in \mathcal{X}} \pi(x)p(x,y). \qquad [4.9]$$

In view of equation [4.1], these are the balance equations [4.2], which thus can be interpreted as the equality between the arrival frequency and the departure frequency in any state at equilibrium.

4.10. Formula of conditional transitions

Consider some stable Markov chain at equilibrium. Let $\mathcal{T} \subset \mathcal{X} \times \mathcal{X}$ be some subset of the transitions of the Markov chain and let $\mathcal{S}$ be some subset of $\mathcal{T}$. We are interested in the probability that, given some transition of the Markov chain in $\mathcal{T}$, this transition belongs to $\mathcal{S}$. Intuitively, this probability corresponds to the ratio of the corresponding frequencies of transitions. At equilibrium, we indeed have:

$$\forall x, y \in \mathcal{X}, \quad P(X_0 = x, X_1 = y) = P(X_0 = x)P(X_1 = y \mid X_0 = x)$$
$$= \pi(x)p(x,y).$$

We deduce:

$$P((X_0, X_1) \in \mathcal{S} \mid (X_0, X_1) \in \mathcal{T}) = \frac{P((X_0, X_1) \in \mathcal{S})}{P((X_0, X_1) \in \mathcal{T})}$$
$$= \frac{\sum_{(x,y) \in \mathcal{S}} \pi(x)p(x,y)}{\sum_{(x,y) \in \mathcal{T}} \pi(x)p(x,y)}$$
$$[4.10]$$

4.11. Chain in reverse time

We have so far defined the Markov chain X_n for all $n \in \mathbb{N}$. Assuming that the chain is stable and at equilibrium, we can build a Markov chain X_n for all $n \in \mathbb{Z}$, representing the evolution of the chain initiated at time $-\infty$, as illustrated in Figure 4.5 for the Markov chain of Figure 4.1.

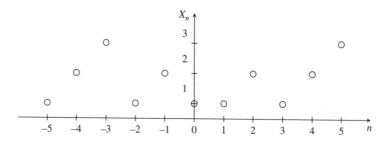

Figure 4.5. *Construction of a Markov chain in reverse time*

We define the chain in *reverse time* $\tilde{X}_n = X_{-n}$, $n \in \mathbb{N}$, whose transition probabilities are given by:

$$\tilde{p}(x,y) = P(\tilde{X}_1 = y \mid \tilde{X}_0 = x),$$

$$= P(X_0 = y \mid X_1 = x),$$

$$= \frac{P(X_0 = y)}{P(X_1 = x)} P(X_1 = x \mid X_0 = y),$$

$$= \frac{\pi(y)}{\pi(x)} p(y,x), \quad \forall x, y \in \mathcal{X}. \qquad [4.11]$$

4.12. Reversibility

A stable chain X_n is said to be *reversible* if it has the same distribution as the associated chain in reverse time $\tilde{X}_n$, that is, if:

$$\forall x, y \in \mathcal{X}, \quad p(x,y) = \tilde{p}(x,y),$$

in view of equation [4.11], we obtain:

$$\forall x, y \in \mathcal{X}, \quad \pi(x)p(x,y) = \pi(y)p(y,x). \qquad [4.12]$$

These are the *local balance* equations, stronger than the global balance equations [4.9], which are obtained by summation. They mean that at equilibrium, the frequency of transition from state x to state y must be equal to the frequency of transition from state y to state x, whatever the states $x, y \in \mathcal{X}$. More generally, we shall say that a Markov chain is reversible if there exists a non-null measure π that verifies the local balance equations [4.12]; in particular, we do not impose stability.

In view of equation [4.3], a necessary condition for reversibility is:

$$p(x, y) > 0 \quad \Longleftrightarrow \quad p(y, x) > 0, \quad \forall x, y \in \mathcal{X}.$$

Any edge in the transition graph must have its symmetrical edge. Such a graph is said to be *symmetric*. In particular, the transition graph of Figure 4.1 is not symmetric and the associated Markov chain is not reversible.

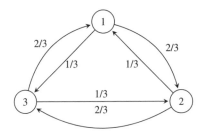

Figure 4.6. *A non-reversible Markov chain with a symmetric transition graph*

REMARK 4.2.– The symmetry of the transition graph is a necessary but not sufficient condition for reversibility. For instance, the Markov chain of Figure 4.6 has a symmetric transition graph but is not reversible: the stationary distribution is $\pi(1) = \pi(2) = \pi(3) = 1/3$ so that the local balance equations [4.12] are violated.

4.13. Kolmogorov's criterion

The following result, due to Kolmogorov[4], gives a criterion for reversibility that can be directly verified on the transition graph. A Markov chain is reversible if and only if its transition graph is symmetric, and for all cycles in this graph, that is for any path $x_0, x_1, \ldots, x_{l-1}, x_0$ of arbitrary length l in the graph, the product of transition probabilities is the same in both directions of the cycle:

$$p(x_0, x_1)p(x_1, x_2) \ldots p(x_{l-1}, x_0)$$
$$= p(x_0, x_{l-1}) \ldots p(x_2, x_1)p(x_1, x_0). \qquad [4.13]$$

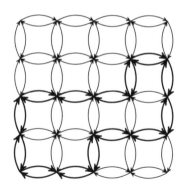

Figure 4.7. *Kolmogorov's criterion: for each cycle in the transition graph, the product of the transition probabilities must be the same in both directions of the cycle*

A solution to the balance equations [4.12] is then given by:

$$\forall x \in \mathcal{X}, \ \pi(x) = \pi(x_0)\frac{p(x_0, x_1)}{p(x_1, x_0)} \frac{p(x_1, x_2)}{p(x_2, x_1)} \ldots \frac{p(x_{l-1}, x)}{p(x, x_{l-1})}, \qquad [4.14]$$

---

4 Andreï Nikolaïevitch Kolmogorov, Russian mathematician (1903–1987).

where $x_0 \in \mathcal{X}$ denotes an arbitrary state and $x_0, x_1, \ldots, x_{l-1}, x$ any path from x_0 to x in the transition graph. This expression is independent of the chosen path: if $x_0, y_1, \ldots, y_{k-1}, x$ denotes any other path from x_0 to x in the transition graph, the equality of the corresponding expressions [4.14] follows from Kolmogorov's criterion [4.13] applied to the cycle $x_0, x_1, \ldots, x_{l-1}, x, y_{k-1}, \ldots, y_1, x_0$. Moreover, we have for any pair of states $x, y \in \mathcal{X}$ linked by an edge in the transition graph:

$$\pi(x)\frac{p(x, y)}{p(y, x)} = \pi(x_0)\frac{p(x_0, x_1)}{p(x_1, x_0)}\frac{p(x_1, x_2)}{p(x_2, x_1)} \cdots \frac{p(x_{l-1}, x)}{p(x, x_{l-1})}\frac{p(x, y)}{p(y, x)}$$
$$= \pi(y),$$

which corresponds to the local balance equations [4.12]. Thus, the Markov chain is reversible.

Conversely, the local balance equations imply the symmetry of the transition graph as well as equality [4.13] since:

$$\frac{p(x_0, x_1)}{p(x_1, x_0)}\frac{p(x_1, x_2)}{p(x_2, x_1)} \cdots \frac{p(x_{l-1}, x_0)}{p(x_0, x_{l-1})} = \frac{\pi(x_1)}{\pi(x_0)}\frac{\pi(x_2)}{\pi(x_1)} \cdots \frac{\pi(x_l)}{\pi(x_{l-1})} = 1.$$

Equation [4.14] shows the interest of reversible Markov chains whose solution to the balance equations is explicit.

In the particular case where the transition graph has a tree structure, in the sense that it is symmetric and the underlying undirected graph is a tree (see Figure 4.8), the Markov chain is always reversible. This is due to the fact that all cycles are symmetric (i.e. of the form $x_0, x_1, \ldots, x_{l-1}, x_l, x_{l-1}, \ldots, x_1, x_0$), so Kolmogorov's criterion is trivially satisfied.

4.14. Truncation of a Markov chain

Let X_n be a Markov chain on $\mathcal{X}$ with stationary measure π. For any non-empty set $A \subset \mathcal{X}$, interpreted as the set of admissible states, we define the truncated chain X'_n as the Markov chain on $\mathcal{A}$ whose transition probabilities from state x to state y, for any $x, y \in \mathcal{A}$, are given by:

$$p'(x,y) = p(x,y), \qquad\qquad\quad \text{if } x \neq y$$
$$p'(x,x) = p(x,x) + \sum_{z \notin \mathcal{A}} p(x,z), \quad \text{otherwise.}$$

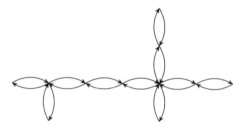

Figure 4.8. *A transition graph with a tree structure*

In other words, any transition from state $x \in \mathcal{A}$ that would lead to a forbidden state (i.e. a state $z \notin \mathcal{A}$) becomes a loop in state x, as illustrated in Figure 4.9. We assume that the truncated chain is irreducible. Figure 4.10 illustrates an example of truncation.

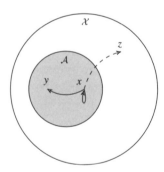

Figure 4.9. *Truncation of a Markov chain*

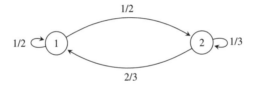

Figure 4.10. *Truncation of the Markov chain of Figure 4.1*
to the subset $\mathcal{A} = \{1,2\}$

When the Markov chain X_n is reversible, the truncated chain X'_n itself is also reversible, with stationary measure π' equal to the restriction of π to $\mathcal{A}$:

$$\pi'(x) = \pi(x), \quad \text{if } x \in \mathcal{A}$$
$$\pi'(x) = 0, \qquad \text{otherwise.}$$

This is a direct consequence of the local balance equations, satisfied by the original Markov chain:

$$\forall x, y \in \mathcal{A}, \quad \pi(x)p(x,y) = \pi(y)p(y,x).$$

When the truncated Markov chain is stable, its stationary distribution is obtained by normalization:

$$\pi'(x) = \frac{\pi(x)}{\sum_{y \in \mathcal{A}} \pi(y)}, \quad \forall x \in \mathcal{A}.$$

4.15. Random walk

We refer to a *random walk* on $\mathcal{X} = \mathbb{N}$ as a Markov chain whose only transitions are those between consecutive integers. We denote by $a(x) = p(x, x+1)$ the probability of walking to the right, and, for all states $x \neq 0$, $b(x) = p(x, x-1)$, the probability of walking to the left, with $a(x) + b(x) = 1$. The associated transition graph is shown in Figure 4.11. This chain, of period 2, is reversible in view of Kolmogorov's criterion, with stationary measure π given by:

$$\pi(x) = \pi(0) \frac{a(1) \dots a(x-1)}{b(1)b(2) \dots b(x)}, \quad \forall x \neq 0.$$

Assuming that the sequences $a(x)$ and $b(x)$ have finite limits given, respectively, by $\bar{a}$ and $\bar{b}$, the random walk is stable if and only if $\bar{a} < \bar{b}$.

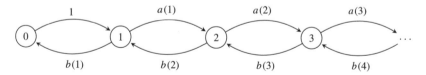

Figure 4.11. *Transition graph of a random walk*

By extension, we also call random walk on $\mathcal{X} = \mathbb{N}$ a Markov chain whose only transitions are those between consecutive integers *and* from an integer to itself, as illustrated in Figure 4.12. We still denote by $a(x) = p(x, x+1)$ the probability of walking to the right, $b(x) = p(x, x-1)$ the probability of walking to the left, and $c(x) = p(x, x)$ the probability of staying at the same place, with $a(x) + b(x) + c(x) = 1$ for all states x.

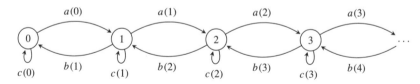

Figure 4.12. *Transition graph of a general random walk*

This chain is aperiodic as soon as $c(x) > 0$ for at least one state x (see Remark 4.1); it is reversible in view of Kolmogorov's criterion, with stationary measure π given by:

$$\pi(x) = \pi(0)\frac{a(0)a(1)\ldots a(x-1)}{b(1)b(2)\ldots b(x)}, \quad \forall x \neq 0.$$

4.16. Exercises

1. First Markov chain

Calculate the stationary distribution of the Markov chain shown in Figure 4.1. Deduce the fraction of time spent in state 1, the frequency of transition from state 1 to state 2, and, given a jump to state 1, the probability that it comes from state 2.

2. Thousand flower honey

A bee gathers pollen in three fields, which it visits in a cyclic manner. In one of the three fields, however, it comes back to the previous field with probability p. Given that it stays the same time in each field before moving to another field, calculate the fraction of time it stays in each field.

3. Traffic information

On some road, three trucks in four are followed by a car and one car in five is followed by a truck. What is the proportion of cars on this road?

4. Distracted runner

A runner lives in a house with two entrance doors. He runs every morning, leaving the house from one of the doors, chosen uniformly at random; when he comes back, he enters the house through one of the doors, also chosen uniformly at random. When he leaves, he takes a pair of shoes at random; when he comes back, he leaves his shoes where he enters the house. If there are no shoes in front of the door he chooses, he runs barefoot. Given that he has C pairs of shoes, what is the probability that he runs barefoot?

5. Walk on a graph

We consider a non-oriented, connex graph without loop from one state to itself. A particle moves at random on this graph. At each move, the particle chooses an edge uniformly at random. Show that the movement of the particle defines a reversible Markov chain and give its stationary distribution.

6. Chessboard

Consider an empty chessboard on which a single piece moves uniformly at random over all admissible movements. Calculate the average number of moves it takes for this piece, starting from any given square, to come back to this square. The different pieces may be considered in the following order: king, tower, bishop, queen, and knight.

4.17. Solution to the exercises

1. First Markov chain

The balance equations [4.2] are given by:

$$\pi(1) = \frac{1}{2}\pi(1) + \frac{2}{3}\pi(2) + \pi(3),$$

$$\pi(2) = \frac{1}{2}\pi(1),$$

$$\pi(3) = \frac{1}{3}\pi(2).$$

Solving this linear system and applying the normalization condition:

$$\pi(1) + \pi(2) + \pi(3) = 1,$$

we get:

$$\pi(1) = \frac{3}{5}, \quad \pi(2) = \frac{3}{10}, \quad \pi(3) = \frac{1}{10}.$$

The fraction of time spent in state 1 is $\pi(1) = 3/5$. The frequency of transition from state 1 to state 2 is $\pi(1)p(1,2) = 3/10$. Given a jump to state 1, the probability that the chain comes from state 2 is the ratio of the corresponding frequencies of transition, that is:

$$\frac{\pi(2)p(2,1)}{\pi(1)} = \frac{1}{3}.$$

2. Thousand flower honey

Let X_n be the field where the bee lies at time n. The choice of the field at time $n+1$ depends only on the field at time n and thus X_n is a Markov chain on $\mathcal{X} = \{1, 2, 3\}$, with the following transition graph:

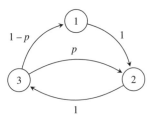

The frequency of visit to each field is given by the stationary distribution π of X_n, see section 4.6. To calculate it, we use the balance equations [4.2]:

$$\pi(1) = (1 - p)\pi(3),$$

$$\pi(2) = p\pi(3) + \pi(1),$$

$$\pi(3) = \pi(2).$$

Using the fact that $\pi(1) + \pi(2) + \pi(3) = 1$, we obtain:

$$\pi(1) = \frac{1-p}{3-p}, \quad \pi(2) = \pi(3) = \frac{1}{3-p}.$$

3. Traffic information

The sequence of vehicles can be described by the Markov chain on $\mathcal{X} = \{T, C\}$ with the following transition graph:

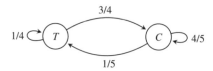

The proportion of cars is then given by $\pi(C)$, where π denotes the stationary distribution of this Markov chain; we obtain $15/19$.

4. Distracted runner

Let X_n be the number of pair of shoes in front of one of the two doors in the morning of day n; X_n is a Markov chain on $\mathcal{X} = \{0, 1, \ldots, C\}$ with the following transition graph:

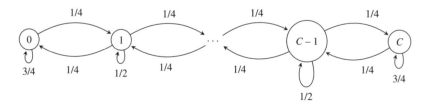

The Markov chain X_n is a random walk, and so the results of section 4.15 can be applied. If π denotes the stationary distribution of X_n then $\pi(x) = \pi(0)$ for all $x \in \{0, \ldots, C\}$. After normalization, we obtain $\pi(0) = 1/(C+1)$. When $X_n \in \{0, C\}$, which occurs with probability $2/(C+1)$, the runner leaves the house barefoot with probability $1/2$; thus, every day he runs barefoot with probability $1/(C+1)$.

5. Walk on a graph

Let X_n be the position of the particle after its nth move. Its position at time $n + 1$ depends only on its position at time n, so X_n is a Markov chain. Since the graph is connex, the chain is irreducible. The transition probability $p(i, j)$ from node i to node j is equal to $1/\deg(i)$, where $\deg(i)$ denotes the degree of node i. Let $\pi(i) = \deg(i)/(2N)$, where N denotes the number of edges of the graph. The probability measure π satisfies the local balance equations:

$$\forall i, j, \quad \pi(i)p(i,j) = \pi(j)p(j,i),$$

so that the Markov chain is reversible, with stationary distribution π.

6. Chessboard

The transition graph of the Markov chain describing the moves of the king on the chessboard is depicted in Figure 4.13, with transition probabilities equal to $1/3$ in the corners (denoted by c hereafter), $1/5$ on the side bands without the corners (denoted by b), and $1/8$ elsewhere (center squares, denoted by a). In view of Kolmogorov's criterion, the Markov chain is reversible. We deduce that the stationary distribution π depends on the type of square a, b, or c, with $\pi(a) = 8/3 \times \pi(c)$ and $\pi(b) = 5/3 \times \pi(c)$. There are 36 squares of type a, 24 of type b, and 4 of type c; the normalization condition is thus:

$$36\pi(a) + 24\pi(b) + 4\pi(c) = 1,$$

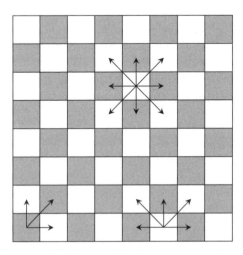

Figure 4.13. *Moves of the king on a chessboard*

from which we deduce:

$$\pi(a) = \frac{2}{105}, \quad \pi(b) = \frac{1}{84}, \quad \pi(c) = \frac{1}{140}.$$

In view of equation [4.7], the king needs on average 52 moves and a half to return to a center square a, starting from this square, 84 moves when starting from a square located on a side band, and 140 when starting from one of the corners.

The method is similar for the other pieces:

– the tower can reach the same number of squares from any square of the chessboard; its position is thus uniformly distributed and the mean return time from each square is equal to 64 moves;

– the bishop can reach 7, 9, 11, and 13 squares from any square located on each of the four concentric rings of the chessboard, from the outer ring to the inner ring; the mean return time is maximum from the outer ring and is then equal to 40 moves;

– the queen can reach 21, 23, 25, and 27 squares from any square located on each of the four concentric rings, from the outer ring to the inner ring; the mean return time is maximum from the outer ring and is then equal to $208/3$ moves;

– the knight can reach 2, 3, 4, 6, or 8 squares depending on its position on the chessboard; the mean return time is maximum from the corners and is then equal to 168 moves. Therefore, the knight is the winner.

Chapter 5

Markov Processes

This chapter describes the definition and main properties of the Markov processes in continuous time. We will see that these may actually be viewed as Markov chains with independent, exponential timers between transitions.

5.1. Definition

We say that a random process $X(t)$, $t \in \mathbb{R}_+$, taking its values in some countable set $\mathcal{X}$, is a Markov process if it is *memoryless* in the sense that for any positive real numbers t, s, any states $x, y \in \mathcal{X}$, any set of l distinct positive real numbers $t_1, \ldots, t_l < t$, and any set of l states $x_1, \ldots, x_l \in \mathcal{X}$:

$$P(X(t+s) = y \mid X(t) = x; X(t_1) = x_1, \ldots, X(t_l) = x_l)$$
$$= P(X(t+s) = y \mid X(t) = x).$$

The Markov process is said to be *homogeneous* if this expression is independent of t.

5.2. Transition rates

In the following sections, we consider only homogeneous and *regular* Markov processes, that is, with an a.s. finite number of transitions over any interval of finite duration. The evolution of such a process, referred to as a *jump process*, is uniquely characterized by its *transition rates* between each pair of states.

For any pair of distinct states $x, y \in \mathcal{X}$, we denote by $q(x, y)$ the transition rate from state x to state y:

$$q(x, y) = \lim_{t \to 0} \frac{1}{t} P(X(t) = y \mid X(0) = x).$$

We assume that for any $x \in \mathcal{X}$, the *departure rate* from state x, defined by:

$$q(x) = \sum_{y \neq x} q(x, y), \qquad \text{[5.1]}$$

satisfies:

$$0 < q(x) < \infty.$$

Using the discrete time, we will see in section 5.3 that the Markov process behaves as follows. When in state x, it virtually moves to state y after some exponential duration with parameter $q(x, y)$ for any $y \neq x$, this duration being infinite if $q(x, y) = 0$. The actual transition is determined by the shortest duration. In view of the properties of the exponential distribution, the transition occurs after some exponential duration with parameter $q(x)$ and corresponds to a change in state y with probability $q(x, y)/q(x)$, independently of this duration.

We refer to the *transition graph* as the oriented graph whose nodes are the elements $\mathcal{X}$ and edges represent transitions of positive rates, as illustrated in Figure 5.1. The weight of each edge is equal to the corresponding transition rate. The process is said to be *irreducible* if its transition graph is strongly connex, that is, if for any distinct states $x, y \in \mathcal{X}$, there exist paths from state x to state y and from state y to state x in the graph. This is what we assume in the following.

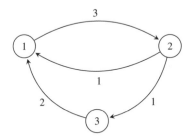

Figure 5.1. *Transition graph of a Markov process on* $\mathcal{X} = \{1, 2, 3\}$

5.3. Discrete analog

For simplicity[1], we assume that the departure rates are bounded:

$$\sup_{x \in \mathcal{X}} q(x) < \infty. \qquad [5.2]$$

We can then define, for any sufficiently small $\tau > 0$, the Markov chain $X_n^{(\tau)}$, $n \in \mathbb{N}$, with values in $\mathcal{X}$ and transition probabilities:

$$p^{(\tau)}(x, y) = \tau q(x, y) \quad \text{if } x \neq y, \qquad [5.3]$$

$$p^{(\tau)}(x, x) = 1 - \tau q(x).$$

The Markov chain $X_n^{(\tau)}$ stays in state x during a geometric random variable with parameter $\tau q(x)$ and then jumps to state

---

1 Exercise 4 in section 5.18 illustrates some issues related to unbounded departure rates.

y with probability $q(x,y)/q(x)$. The stochastic process equal to $X_n^{(\tau)}$ on the time interval $[n\tau, (n+1)\tau)$ behaves like the Markov process $X(t)$ when τ tends to zero: in view of section 2.2, the limiting process stays in state x during an exponential duration with parameter $q(x)$ and then jumps to state y with probability $q(x,y)/q(x)$.

We refer to $X_n^{(\tau)}$ as the *skeleton chain* associated with the Markov process $X(t)$. This is the discrete time version of the process. Thus all properties of Markov chains apply to Markov processes. The only exception is the notion of periodicity, which is inherent to discrete time. Note that the skeleton chain is aperiodic since $p^{(\tau)}(x,x) > 0$ for all states x (see Remark 4.1): a Markov process cannot be periodic.

REMARK 5.1.– (Sampling). The skeleton chain $X_n^{(\tau)}$ is *not* the Markov process $X(t)$ sampled at times $n\tau$, $n \in \mathbb{N}$. In particular, the Markov process may have an arbitrarily large number of jumps in any interval of duration τ. The skeleton chain may be viewed as the sampled Markov process at times $n\tau$ in the limit $\tau \to 0$ only.

5.4. Balance equations

We say that a probability measure π on $\mathcal{X}$ is a *stationary distribution* of the process if the distribution of $X(t)$ is equal to π at any time $t \geq 0$, whenever the distribution of the initial state $X(0)$ is equal to π. We then say that the process is in a *steady state*.

Any stationary distribution satisfies the equations:

$$\forall x \in \mathcal{X}, \quad \pi(x) \sum_{y \neq x} q(x,y) = \sum_{y \neq x} \pi(y) q(y,x). \qquad [5.4]$$

These are *balance equations* of the Markov process. They follow from the skeleton chain $X_n^{(\tau)}$. Any stationary distribution $\pi^{(\tau)}$ of this chain indeed satisfies the corresponding balance equations:

$$\forall x \in \mathcal{X}, \quad \pi^{(\tau)}(x) = \sum_{y \in \mathcal{X}} \pi^{(\tau)}(y) p^{(\tau)}(y, x).$$

In view of equation [5.3], these equations are equivalent to:

$$\forall x \in \mathcal{X}, \quad \pi^{(\tau)}(x) \sum_{y \neq x} q(x, y) = \sum_{y \neq x} \pi^{(\tau)}(y) q(y, x),$$

and thus coincide with the balance equations of the Markov process [5.4].

5.5. Stationary measure

We refer to a *stationary measure* as any non-null solution to the balance equations; a stationary measure is defined up to some multiplicative constant and is thus generally not a probability measure. As for the Markov chains, it follows from irreducibility that any stationary measure satisfies:

$$\forall x \in \mathcal{X}, \quad \pi(x) > 0. \tag{5.5}$$

There is a *stationary distribution* if and only if there is a stationary measure of finite sum. The stationary distribution π is then unique and follows from the normalization of the stationary measure, so that:

$$\sum_{x \in \mathcal{X}} \pi(x) = 1. \tag{5.6}$$

5.6. Stability and ergodicity

A Markov process having some stationary distribution π is said to be *stable*. This stationary distribution is also the limiting distribution of the process in the sense that, for any state x:

$$\lim_{t \to \infty} \mathrm{P}(X(t) = x) = \pi(x),$$

and this for any initial state $X(0)$. Moreover, the stationary distribution gives the fraction of time spent in each state, that is, for any state x:

$$\lim_{T \to \infty} \frac{1}{T} \int_0^T 1(X(t) = x)\, \mathrm{d}t = \pi(x) \quad \text{a.s.}$$

whatever the initial state $X(0)$. This is illustrated in Figure 5.2.

More generally, we have for any function $f \colon \mathcal{X} \to \mathbb{R}$ such that $\mathrm{E}(f(X)) < \infty$:

$$\lim_{T \to \infty} \frac{1}{T} \int_0^T f(X(t))\, \mathrm{d}t = \mathrm{E}(f(X)) \quad \text{a.s.} \qquad [5.7]$$

where X is a random variable with distribution π. This is the *ergodic theorem*, which states that the means in time (the left-hand side of the equality) and in space (the right-hand side of the equality) coincide. The process is said to be *ergodic*.

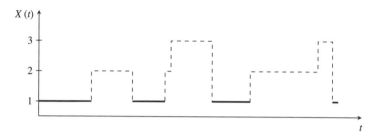

Figure 5.2. *Ergodicity: π(1) is the fraction of time spent in state 1*

A Markov process that does not have any stationary distribution is said to be *unstable*. It then goes to infinity in the sense that, for any state x:

$$\lim_{t \to \infty} P(X(t) = x) = 0,$$

whatever the initial state $X(0)$. Thus the process eventually leaves any finite subset of states. In particular, if the state space is finite, the process is necessarily stable. For an infinite state space, the process is unstable if there is no solution of finite sum to the balance equations [5.4].

5.7. Recurrence and transience

Like Markov chains, the standard classification of the Markov processes is based on the notions of recurrence and transience: a process is said to be *recurrent* if the return time to any state, that is, the time it takes to leave the state and come back to this state, is a.s. finite, and *transient* otherwise. A recurrent process is said to be *positive* if this return time has a finite mean, and *null* otherwise. According to our terminology, a stable process is positive recurrent whereas an unstable process is either null recurrent or transient. For a stable process, the mean return time to any state x is the inverse of the frequency of jumps to this state, that is, in view of section 5.8:

$$\frac{1}{\pi(x)q(x)}.$$

5.8. Frequency of transition

For any pair of distinct states $x, y \in \mathcal{X}$, the quantity $\pi(x)q(x, y)$ corresponds to the *frequency of transition* from state x to state y, as illustrated in Figure 5.3. This follows from the application of equation [4.8] to the skeleton chain, for any sufficiently small τ:

$$\pi(x)q(x,y) = \lim_{N\to\infty} \frac{1}{N\tau} \sum_{n=1}^{N} 1\left(X_n^{(\tau)} = x, X_{n+1}^{(\tau)} = y\right) \quad \text{a.s.}$$

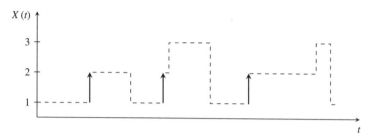

Figure 5.3. *Frequency of transition from state 1 to state 2, given by π(1)q(1, 2)*

In particular, the arrival frequency to state x and the departure frequency from state x are, respectively, given by:

$$\sum_{y\neq x} \pi(y)q(y,x) \quad \text{and} \quad \sum_{y\neq x} \pi(x)q(x,y).$$

These quantities must be equal at the equilibrium. We find the balance equations [5.4], which may indeed be interpreted as the equality of arrival frequency and departure frequency in any state. Note that, in view of equation [5.1], the frequency of jumps to state x is simply given by:

$$\pi(x)q(x).$$

5.9. Virtual transitions

It is useful to consider, in addition to the transitions between the distinct states considered so far, the *virtual* transitions of the Markov process from one state to itself.

For any state $x \in \mathcal{X}$, we denote by $q(x, x)$, the transition rate from state x to itself. The transition graph now has loops in all states x such that $q(x, x) > 0$, as illustrated in Figure 5.5. The departure rate from any state $x \in \mathcal{X}$ is now defined by:

$$q(x) = \sum_{y \in \mathcal{X}} q(x, y).$$

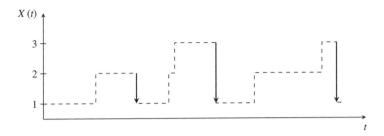

Figure 5.4. *Frequency of jumps to state 1, given by* $\pi(1)q(1)$

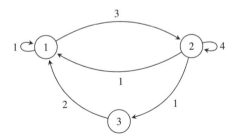

Figure 5.5. *Transition graph of a Markov process with virtual transitions*

The virtual transitions do not change the evolution of the Markov process but introduce additional events, namely jumps of the process from one state to itself. This is illustrated in Figure 5.6 for the Markov process shown in Figure 5.5. When the process is in state x, it virtually jumps to state y after some exponential duration of parameter $q(x, y)$; for any $y \in \mathcal{X}$, this duration being infinite when $q(x, y) = 0$. The actual change of state, determined by the shortest duration, occurs after some exponential duration of parameter $q(x)$ and corresponds to a jump to state y with probability $q(x, y)/q(x)$, independently of this duration, for any $y \in \mathcal{X}$.

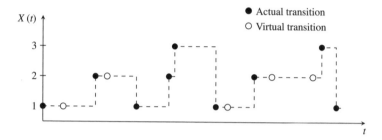

Figure 5.6. *Evolution of a Markov process with virtual transitions*

The resulting Markov process has the same distribution as a Markov process with the same transition rates but without virtual transitions. The balance equations are given by:

$$\pi(x) \sum_{y \in \mathcal{X}} q(x, y) = \sum_{y \in \mathcal{X}} \pi(y) q(y, x), \quad \forall x \in \mathcal{X}. \qquad [5.8]$$

These equations are equivalent to the balance equations without virtual transitions, equation [5.4].

The quantity $\pi(x) q(x)$ corresponds to the frequency of jumps to state x, including the virtual transitions, while its inverse $1/(\pi(x) q(x))$ corresponds to the mean return time to state x, the return being possibly due to a virtual transition.

5.10. Embedded chain

Let $X_0 = X(0)$ and X_n be the state of the Markov process $X(t)$ just after its nth jump, possibly including virtual transitions. The sequence X_n, $n \in \mathbb{N}$, defines a Markov chain, called the *embedded chain*, whose transition probabilities are given by:

$$\forall x, y \in \mathcal{X}, \quad p(x,y) = \frac{q(x,y)}{q(x)}. \qquad [5.9]$$

If π is a stationary measure of the process, then the measure π^0 defined by:

$$\forall x \in \mathcal{X}, \quad \pi^0(x) = \pi(x)q(x), \qquad [5.10]$$

is a stationary measure of the embedded chain:

$$\forall x \in \mathcal{X}, \quad \pi^0(x) = \sum_{y \in \mathcal{X}} \pi^0(y)p(y,x).$$

The irreducibility of the Markov process (and thus of the embedded chain) implies that any solution to these equations satisfies $\pi^0(x) > 0$ for any $x \in \mathcal{X}$.

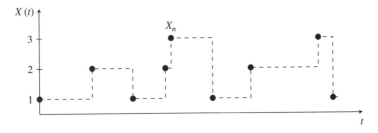

Figure 5.7. *A Markov process and its embedded Markov chain*

In view of equation [5.2], the Markov process is stable if and only if its embedded chain is stable. Denoting by π and π^0 their respective stationary distributions, the probability $\pi^0(x)$ that the embedded chain is in state x is proportional to $\pi(x)q(x)$, the frequency of the visit of state x. Whereas the stationary distribution of the process, π, gives the proportion of *time* the process is in each state (see the ergodic theorem in section 5.6), the stationary distribution of the embedded chain, π^0, gives the proportion of *jumps* of the process to each state. The embedded chain may be viewed as a Markov process whose exponential timers have been homogenized, so that the process spends the same mean time in each state before the next transition.

5.11. Formula of conditional transitions

Let $\mathcal{T} \subset \mathcal{X} \times \mathcal{X}$ be a subset of the transitions of the Markov process and $\mathcal{S}$ be a subset of $\mathcal{T}$. We consider the event that, given some transition in $\mathcal{T}$, this transition belongs to $\mathcal{S}$. This probability is equal to the ratio of the frequencies of the corresponding transitions. The embedded chain allows us to prove this result. It is sufficient to observe that, in a steady state:

$$\forall x, y \in \mathcal{X}, \quad P(X_0 = x, X_1 = y),$$
$$= P(X_0 = x)P(X_1 = y \mid X_0 = x),$$
$$= \pi^0(x)p(x,y).$$

In view of equations [5.9] and [5.10], this quantity is proportional to $\pi(x)q(x,y)$, the frequency of transition from state x to state y. We deduce:

$$P((X_0, X_1) \in \mathcal{S} \mid (X_0, X_1) \in \mathcal{T}) = \frac{P((X_0, X_1) \in \mathcal{S})}{P((X_0, X_1) \in \mathcal{T})},$$
$$= \frac{\sum_{(x,y) \in \mathcal{S}} \pi(x)q(x,y)}{\sum_{(x,y) \in \mathcal{T}} \pi(x)q(x,y)}.$$

5.12. Process in reverse time

So far we have defined the process $X(t)$ for any $t \in \mathbb{R}_+$. Assuming that the distribution of $X(0)$ is the stationary distribution π, the Markov process $X(t)$ can be extended to any $t \in \mathbb{R}$. For this purpose, we define the process in reverse time $\tilde{X}(t) = X(-t)$, $t \in \mathbb{R}_+$, with the same departure rates as those of the process $X(t)$ and whose transition probabilities of the embedded chain $\tilde{X}_n$ are given by:

$$
\begin{aligned}
\forall x, y \in \mathcal{X}, \quad \tilde{p}(x,y) &= \mathrm{P}(\tilde{X}_1 = y \mid \tilde{X}_0 = x), \\
&= \mathrm{P}(X_0 = y \mid X_1 = x), \\
&= \frac{\mathrm{P}(X_0 = y)}{\mathrm{P}(X_1 = x)} \mathrm{P}(X_1 = x \mid X_0 = y), \\
&= \frac{\pi^0(y)}{\pi^0(x)} p(y, x).
\end{aligned}
$$

assuming the embedded chain X_n is stable and in steady state.

Using equations [5.9] and [5.10], we obtain the transition rates of the process $\tilde{X}(t)$:

$$
\begin{aligned}
\forall x, y \in \mathcal{X}, \quad \tilde{q}(x,y) &= \tilde{p}(x,y) q(x), \\
&= \frac{\pi(y) q(y)}{\pi(x) q(x)} p(y, x) q(x), \\
&= \frac{\pi(y)}{\pi(x)} q(y, x). \qquad\qquad [5.11]
\end{aligned}
$$

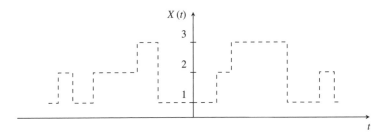

Figure 5.8. *Construction of a Markov process in reverse time*

5.13. Reversibility

The process $X(t)$ is said to be *reversible* if it has the same distribution as the process in reverse time $\tilde{X}(t)$, that is, if:

$$\forall x, y \in \mathcal{X}, \quad \tilde{q}(x, y) = q(x, y).$$

In view of equation [5.11], we get:

$$\pi(x)q(x, y) = \pi(y)q(y, x), \quad \forall x, y \in \mathcal{X}. \tag{5.12}$$

These are the *local balance* equations, stronger than the global balance equations [5.8], which are obtained by summation. It means that in a steady state, the frequency of transition from state x to state y must be equal to that of the transition from state y to state x, for all states $x, y \in \mathcal{X}$. More generally, we shall say that a Markov process is reversible if there exists a non-null measure π that verifies the local balance equations [5.12]; we do not impose stability.

In view of equation [5.5], a necessary (but not sufficient) condition for reversibility is:

$$q(x, y) > 0 \quad \Longleftrightarrow \quad q(y, x) > 0, \quad \forall x, y \in \mathcal{X}.$$

Any edge in the transition graph must have its symmetrical edge. Such a graph will be said to be *symmetric*. For instance, the transition graph represented in Figure 5.1 is not symmetric and thus does not correspond to a reversible Markov process.

5.14. Kolmogorov's criterion

A Markov process is reversible if and only if its transition graph is symmetric and for any cycle in this graph, that is, any path $x_0, x_1, \ldots, x_{l-1}, x_0$ of arbitrary length l in the graph, the product of the transition rates is the same in both directions of the cycle:

$$q(x_0, x_1)q(x_1, x_2) \ldots q(x_{l-1}, x_0)$$
$$= q(x_0, x_{l-1}) \ldots q(x_2, x_1)q(x_1, x_0). \qquad [5.13]$$

A solution to the balance equations [5.12] is then given by:

$$\forall x \in \mathcal{X}, \quad \pi(x) = \pi(x_0)\frac{q(x_0, x_1)}{q(x_1, x_0)} \frac{q(x_1, x_2)}{q(x_2, x_1)} \ldots \frac{q(x_{l-1}, x)}{q(x, x_{l-1})}, \quad [5.14]$$

where x_0 is an arbitrary state and $x_0, x_1, \ldots, x_{l-1}, x$ denotes any path from x_0 to x in the transition graph. This expression is independent of the chosen path and we indeed have, for any pair of states $x, y \in \mathcal{X}$ connected by an edge in the transition graph:

$$\pi(x)\frac{q(x, y)}{q(y, x)} = \pi(x_0)\frac{q(x_0, x_1)}{q(x_1, x_0)} \frac{q(x_1, x_2)}{q(x_2, x_1)} \ldots \frac{q(x_{l-1}, x)}{q(x, x_{l-1})} \frac{q(x, y)}{q(y, x)} = \pi(y),$$

which corresponds to the local balance equations [5.12].

Conversely, the local balance equations imply the symmetry of the transition graph as well as equality [5.13] since:

$$\frac{q(x_0, x_1)}{q(x_1, x_0)} \frac{q(x_1, x_2)}{q(x_2, x_1)} \ldots \frac{q(x_{l-1}, x)}{q(x, x_{l-1})} = \frac{\pi(x_1)}{\pi(x_0)} \frac{\pi(x_2)}{\pi(x_1)} \ldots \frac{\pi(x_l)}{\pi(x_{l-1})} = 1.$$

Equation [5.14] shows the interest of reversible processes, whose solution to the balance equations is explicit.

5.15. Truncation of a reversible process

Let $X(t)$ be a Markov process on $\mathcal{X}$ with a stationary measure π. For any non-empty subset $\mathcal{A} \subset \mathcal{X}$ of the state space, we define the truncated process $X'(t)$ as the Markov process on the state space $\mathcal{A}$ whose transition rates are given by:

$$q'(x,y) = q(x,y), \quad \forall x,y \in \mathcal{A}.$$

Thus any transition from state $x \in \mathcal{A}$ that would lead to a forbidden state (i.e. a state $z \notin \mathcal{A}$) is removed, as illustrated in Figure 5.9. Figure 5.10 gives an example of such a truncation. We assume that the truncated process is irreducible.

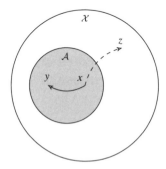

Figure 5.9. *Truncation of a Markov process*

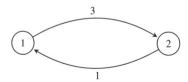

Figure 5.10. *Truncation of the Markov process of Figure 5.1 to the subset $\mathcal{A} = \{1, 2\}$*

When the Markov process $X(t)$ is reversible, the truncated process $X'(t)$ is itself reversible, with stationary measure π' equal to the restriction of π to $\mathcal{A}$:

$$\pi'(x) = \pi(x) \quad \text{if } x \in \mathcal{A},$$
$$\pi'(x) = 0 \quad \quad \text{otherwise.}$$

This is a consequence of the local balance equations, satisfied by the original Markov process:

$$\forall x, y \in \mathcal{A}, \quad \pi(x)q(x,y) = \pi(y)q(y,x).$$

When the truncated Markov process is stable, its stationary distribution is obtained by normalization:

$$\pi'(x) = \frac{\pi(x)}{\sum_{y \in \mathcal{A}} \pi(y)}, \quad \forall x \in \mathcal{A}.$$

5.16. Product of independent Markov processes

Let $X_1(t)$ and $X_2(t)$ be two independent Markov processes on respective state spaces $\mathcal{X}_1$ and $\mathcal{X}_2$ and with respective stationary measures π_1 and π_2. We denote by $q_1(x_1, y_1)$, $q_2(x_2, y_2)$ the corresponding transition rates without any virtual transition. The product process $X(t) = (X_1(t), X_2(t))$ is a Markov process on the state space $\mathcal{X} = (\mathcal{X}_1, \mathcal{X}_2)$, with transition rates:

$$q(x,y) = \begin{cases} q_1(x_1, y_1) & \text{if } x_2 = y_2, \\ q_2(x_2, y_2) & \text{if } x_1 = y_1, \\ 0 & \text{otherwise} \end{cases} \quad \forall x, y \in \mathcal{X},$$

and stationary measure:

$$\pi(x) = \pi_1(x_1)\pi_2(x_2), \quad \forall x \in \mathcal{X}.$$

Moreover, the process $X(t)$ is reversible whenever $X_1(t)$ and $X_2(t)$ are reversible. For all $x, y \in X$ such that $x_2 = y_2$ for instance, we indeed obtain:

$$\begin{aligned} \pi(x)q(x,y) &= \pi_1(x_1)\pi_2(x_2)q_1(x_1, y_1), \\ &= \pi_1(y_1)\pi_2(x_2)q_1(y_1, x_1), \\ &= \pi(y)q(y,x). \end{aligned}$$

5.17. Birth–death processes

We refer to a birth–death process as a Markov process on $\mathcal{X} = \mathbb{N}$ whose transitions occur between consecutive integers only. For any $x \in \mathbb{N}$, we denote by $\lambda(x) = q(x, x+1)$ the birth rate in state x and $\mu(x) = q(x, x-1)$ the death rate in state x, for any $x > 0$. The transition graph of this process is represented in Figure 5.11. Such a process is reversible in view of Kolmogorov's criterion, with stationary measure π given by:

$$\pi(x) = \pi(0)\frac{\lambda(0)\lambda(1)\ldots\lambda(x-1)}{\mu(1)\mu(2)\ldots\mu(x)}, \quad \forall x \neq 0 \qquad [5.15]$$

Assuming that the birth–death rates $\lambda(x)$ and $\mu(x)$ have finite limits, denoted by $\bar{\lambda}$ and $\bar{\mu}$, respectively the process is stable if and only if $\bar{\lambda} < \bar{\mu}$.

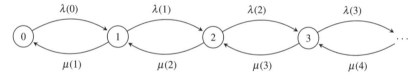

Figure 5.11. *Transition graph of a birth–death process*

5.18. Exercises

1. First Markov process

Calculate the stationary distribution of the Markov process in Figure 5.1. Deduce the fraction of time spent in state 1 and, given a jump to state 1, the probability that the process comes from state 2. Compare it with the results obtained for the associated embedded Markov chain.

2. *Walk on a cube*

We consider a Markov process on the state space formed by the vertices of a cube. The transitions that are allowed are along the cube edges as well as one of the four large diagonals of the cube. The transition rates are all equal to 1. Calculate the frequency of passage through the diagonal at equilibrium.

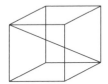

3. *Parking space*

We consider two parking spaces, one private reserved for subscribers, another public and accessible to all. The subscribers arrive according to a Poisson process of intensity λ_1 and, when both spaces are available, choose one uniformly at random. The non-subscribers arrive according to a Poisson process of intensity λ_2 and can only access the public parking space. A car that does not find any available parking space goes away. Each car is parked during an exponential duration with parameter μ.

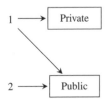

We describe the system through the state of each parking space, available or occupied. Draw the transition graph of the associated Markov process. Is this process reversible? Give the probability that a subscriber does not find any available parking space as a function of the stationary distribution of the process. Answer the same question for a non-subscriber. Calculate the numerical values when $\lambda_1 = \lambda_2 = \mu = 1$.

4.   Balloon drop

A balloon, subject to wind speed, has an erratic trajectory. It stays at a height of x meters during an exponential duration then goes up to $x+1$ meters or down to $x-1$ meters with equal probability (except when $x = 0$, where it always goes up to $x = 1$ meter). Does it eventually take off when the mean time elapsed at height x before each move is equal to 1, $1/(x + 1)$, and $1/(x + 1)^2$? Otherwise, which fraction of time does it stay on the ground?

5.   Partition

Consider a stable Markov process on some state space $\mathcal{X}$. Given some partition of the set $\mathcal{X}$ into two parts $\mathcal{A}$ and $\mathcal{B}$, show that at equilibrium, the frequency of the passage of the process from part $\mathcal{A}$ to part $\mathcal{B}$ is equal to the frequency of the passage from part $\mathcal{B}$ to part $\mathcal{A}$.

6.   Guided tour

A castle consists of N successive rooms, indexed from 1 to N in the direction of the visit. Three guides share the visits. It takes 4 min on average for a single guide to describe the room; the duration is random and exponentially distributed. Since the guides cannot talk simultaneously in the same room, a guide must wait for the end of the talk of the previous guide before starting her talk. Calculate the average duration of the visit, given that guides do not stop working. We might describe the system state by the number of guides in each room and verify that the stationary distribution is uniform on the set of states.

7.   The conquest of the West
A territory is divided into 16 regions as follows:

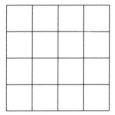

A gold-seeker moves in this territory between neighboring regions according to some Markov process: its transition rates are equal to 1, 2, 1, 3 toward the north, the south, the east, and the west, respectively. In which region is the gold-seeker more likely to stay? What is the associated probability?

Two gold-seekers must now share the territory. The transition rates among the 16 regions are the same, except that the two gold-seekers cannot be simultaneously in the same region. What is the probability that both gold-seekers lie in the western half of the territory?

8.   Bike rental
Consider N bike stations consisting of $K_1, \ldots, K_N$ parking spaces. Customers arrive at these stations according to independent Poisson processes with respective intensities $\lambda_1, \ldots, \lambda_N$. A customer that does not find any available bike goes away. Otherwise, he or she takes a bike for some exponential duration of parameter μ and then attempts to return it at station i with probability p_i, with $p_1 + \ldots + p_N = 1$. If no parking space is available, the customer goes away for a new exponential duration of parameter μ, before attempting to return the bike in one of the stations, chosen again with probabilities $p_1, \ldots, p_N$.

Assuming that the stations are initially full, calculate the probability that a customer arriving at some station does not find any available bike, and the probability that a customer attempting to return his or her bike cannot do so. Calculate the numerical values for $N = 2$ stations with two parking spaces each, with $\lambda_1 = \lambda_2 = \mu = 1$ and $p_1 = p_2 = 1/2$.

5.19. Solution to the exercises

1. *First Markov process*
   The balance equations [5.4] are given by:

$$3\pi(1) = \pi(2) + 2\pi(3),$$

$$2\pi(2) = 3\pi(1),$$

$$2\pi(3) = \pi(2).$$

Solving this linear system and applying the normalization condition:

$$\pi(1) + \pi(2) + \pi(3) = 1,$$

we get:

$$\pi(1) = \frac{4}{13}, \quad \pi(2) = \frac{6}{13}, \quad \pi(3) = \frac{3}{13}.$$

The fraction of time spent in state 1 is $\pi(1) = 4/13$. Given a jump to state 1, the probability that the process comes from state 2 is the ratio of the corresponding frequencies of transition, that is:

$$\frac{\pi(2)q(2,1)}{\pi(1)q(1)} = \frac{1}{2}.$$

The embedded Markov chain has the following transition graph:

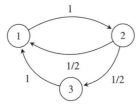

Using equation [5.10] and applying the normalization condition:

$$\pi^0(1) + \pi^0(2) + \pi^0(3) = 1,$$

we get:

$$\pi^0(1) = \frac{2}{5}, \quad \pi^0(2) = \frac{2}{5}, \quad \pi^0(3) = \frac{1}{5}.$$

The fraction of jumps to state 1 is $\pi^0(1) = 2/5$, which is different from that of time spent by the Markov process in state 1. Given a jump to state 1, the probability that the Markov chain comes from state 2 is the ratio of the corresponding frequencies of transition, that is:

$$\frac{\pi^0(2)p(2,1)}{\pi^0(1)} = \frac{1}{2}$$

Of course, this value corresponds to that obtained above with the Markov process.

2. *Walk on a cube*

Using Kolmogorov's criterion, we verify that the Markov process is reversible and that its stationary distribution is uniform over all vertices. The frequency of transition through the diagonal is the sum of the transition frequencies in both directions, that is $1/4$.

3.   *Parking space*

We denote by $X(t)$ the state of the private parking space at time t: $X(t)$ is equal to 0 if the space is free and to 1 if the space is occupied. In the same manner, we denote by $Y(t)$ the state of the public parking space at time t. The joint process $(X(t), Y(t))$ is a Markov process whose transition graph is the following:

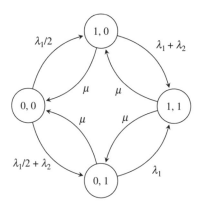

Applying Kolmogorov's criterion to the cycle $(0,0) \rightarrow (1,0) \rightarrow (1,1) \rightarrow (0,1) \rightarrow (0,0)$, we verify that this Markov process is not reversible. We must use the balance equations to derive the stationary distribution π. From the Poisson arrivals see time averages (PASTA) property, the probabilities that a subscriber and a non-subscriber will not find any available parking space are, respectively, given by $p_1 = \pi(1,1)$ and $p_2 = \pi(0,1) + \pi(1,1)$.

In the specific case where $\lambda_1 = \lambda_2 = \mu = 1$, we obtain:

$$\pi(0,0) = \frac{10}{43}, \quad \pi(1,0) = \frac{6}{43}, \quad \pi(0,1) = \frac{14}{43}, \quad \pi(1,1) = \frac{13}{43}.$$

We deduce that $p_1 = 13/43 \approx 0.30$ and $p_2 = 27/43 \approx 0.62$

4. *Balloon drop*

We denote by $X(t)$ the height of the balloon at time t. The balloon stays at height x during the exponential time and then moves to a different height that depends on x only: $X(t)$ is a Markov process whose transition graph is the following, where $a(x)$ is the parameter of the exponential distribution associated with the time spent at height x:

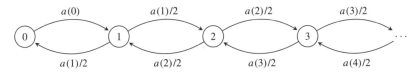

This is a birth–death process whose stationary measure is given by:

$$\pi(x) = \pi(0)\frac{2a(0)}{a(x)}, \ \forall x \geq 1$$

The process is stable if and only if π is summable. This is not the case if $a(x) = 1$ or $a(x) = x + 1$. If $a(x) = (x + 1)^2$, the process is stable; after normalization, we obtain the fraction of time spent on the ground: $\pi(0) = 3/(\pi^2 - 3) \approx 0.44$.

NOTE: The embedded Markov chain is the same in the three cases and is unstable. When the transition rates are not bounded, there is no simple relationship between the stability of the process and the stability of the embedded chain. Following the same principle, the reader might build an unstable Markov process whose embedded chain is stable.

5. *Partition*

The balance equations [5.4] can be written:

$$\forall x \in \mathcal{X}, \quad \pi(x)\left(\sum_{y\in\mathcal{A}} q(x,y) + \sum_{y\in\mathcal{B}} q(x,y)\right),$$

$$= \sum_{y\in\mathcal{A}} \pi(y)q(y,x) + \sum_{y\in\mathcal{B}} \pi(y)q(y,x).$$

Summing these equations over all $x \in \mathcal{A}$ and simplifying the resulting expression, we obtain the desired result:

$$\sum_{x \in \mathcal{A}}\sum_{y \in \mathcal{B}} \pi(x)q(x,y) = \sum_{x \in \mathcal{A}}\sum_{y \in \mathcal{B}} \pi(y)q(y,x).$$

6.   *Guided tour*

The number of guides in each room forms a Markov process on the state space:

$$\mathcal{X} = \{x \in \mathbb{N}^N : x_1 + \ldots + x_N = 3\}.$$

Denoting by e_i the unit vector associated with the ith component, the transition rate of the process from state x to state $x - e_i + e_{i+1}$ is equal to μ for any $i = 1, 2, \ldots, N - 1$ and any x such that $x_i \neq 0$, where μ is the departure rate of an active guide from a room, that is, $\mu = 1/4$ min^{-1}; moreover, given that the guides do not stop working, the transition rate from state x to state $x - e_N + e_1$ is equal to μ for any x such that $x_N \neq 0$; the other transition rates are null.

NOTE: The process is reversible only for $N = 2$. For $N = 3$, for instance, there is a transition from state $(3,0,0)$ to state $(2,1,0)$ but not in the other direction.

The balance equations are:

$$\forall x \in \mathcal{X}, \quad \pi(x) \sum_{i:x_i \neq 0} \mu = \sum_{i:x_i \neq 0} \pi(x - e_i + e_{i-1})\mu,$$

with the convention $e_0 \equiv e_N$. We verify that the measure π given by $\pi(x) = 1$ for any $x \in \mathcal{X}$ is stationary. By normalization, the stationary distribution is thus given by $\pi(x) = 1/\Gamma_N$ for any $x \in \mathcal{X}$, where Γ_N is the number of states, that is, considering the respective cases in which three guides are in the same room, two guides are in the same room, and three guides are in distinct rooms:

$$\Gamma_N = N + N(N-1) + \frac{N(N-1)(N-2)}{3!} = \frac{N(N+1)(N+2)}{6}.$$

To conclude, the departure frequency of guides from any room is given by the product of μ by the probability that this room is occupied, that is:

$$\left(1 - \frac{\Gamma_{N-1}}{\Gamma_N}\right) \times \mu = \frac{3\mu}{N+2}.$$

We deduce the arrival frequency of a guide in any room:

$$\frac{\mu}{N+2}$$

and the mean guided tour duration:

$$\frac{N+2}{\mu}.$$

The average waiting time due to the other guides is thus equal to $2/\mu$, that is, $8\,\mathrm{min}$, whatever the number of rooms.

7.   The conquest of the West

Let $X(t)$ be the region explored by the gold-seeker at time t. Using Kolmogorov's criterion, we check that this process is reversible. Indexing the regions from 0 to 3 from north to south and from east to west, we obtain the stationary measure:

$$\pi(i,j) = 3^i 2^j \pi(0,0), \quad \forall i,j \in \{0,\ldots,3\}.$$

The normalization gives:

$$\pi(0,0) = \frac{1}{(1+3+9+27) \times (1+2+4+8)} = \frac{1}{600}.$$

Thus, it is more likely to find the gold-seeker in region $(3,3;$ south-west) and this happens with a probability $9/25$.

Let $X_1(t)$ and $X_2(t)$ be the locations of both gold-seekers at time t. We first consider the joint process $(X_1(t), X_2(t))$ without any constraint (i.e. both gold-seekers can explore the same region simultaneously). This is a reversible process, the product of two independent reversible processes. To add the constraints, it is then sufficient to consider the truncation of this process (see section 5.15); the probability that both gold-seekers lie in the western half of the territory is given by:

$$\frac{\sum_{(i_1,j_1)\neq(i_2,j_2),j_1\geq 2,j_2\geq 2} \pi(i_1,j_1)\pi(i_2,j_2)}{\sum_{(i_1,j_1)\neq(i_2,j_2)} \pi(i_1,j_1)\pi(i_2,j_2)} = \frac{4455}{5806} \approx 0.77.$$

8. *Bike rental*

The numbers of free spaces in the N stations form a Markov process on the state space:

$$\mathcal{X} = \{0,1,\ldots,K_1\} \times \ldots \times \{0,1,\ldots,K_N\}.$$

Denoting by e_i the unit vector on component i and defining $\mu_i = \mu p_i$, the transition rates of the process are equal to λ_i from state x and state $x + e_i$, for any x such that $x_i < K_i$, and to $|x|\mu_i$ from state x to state $x - e_i$, for any x such that $x_i > 0$, where $|x| = x_1 + x_2 + \ldots + x_N$ denotes the total number of borrowed bikes. We verify using Kolmogorov's criterion that this Markov process is reversible and has the following stationary measure:

$$\pi(x) = \frac{\lambda_1^{x_1}\ldots\lambda_N^{x_N}}{|x|!\mu_1^{x_1}\ldots\mu_N^{x_N}}, \quad \forall x \in \mathcal{X}.$$

The stationary distribution is obtained by normalization. To calculate the probability that a customer arriving at some station does not find any available bike, we introduce virtual jumps of rate λ_i in any state x such that $x_i = K_i$. The probability is then the ratio of the associated transition frequencies, that is:

$$p = \frac{\sum_{i=1}^{N} \sum_{x:x_i=K_i} \pi(x)\lambda_i}{\sum_{i=1}^{N} \lambda_i}.$$

Similarly, the probability that a customer attempting to return her bike cannot do so is given by:

$$q = \frac{\sum_{i=1}^{N} \sum_{x:x_i=0} \pi(x)|x|\mu_i}{\sum_x \pi(x)|x|\mu}.$$

For $N = 2$ stations of two parking spaces each, with $\lambda_1 = \lambda_2 = \mu = 1$ and $p_1 = p_2 = 1/2$, we find:

$$p = \frac{12}{43} \approx 0.28 \quad \text{and} \quad q = \frac{9}{40} \approx 0.22.$$

Chapter 6

Queues

In the following two chapters, we introduce the key elements of queueing theory, which rely on the Markov processes described in the previous chapter and are instrumental in the analysis of subsequent traffic models.

6.1. Kendall's notation

A queue is characterized by several parameters such as the number of servers, the queue capacity in the number of customers, and the statistical characteristics of customer arrivals and service times. For convenience, the following code invented by Kendall[1] is commonly used:

$$A/S/m[/n]$$

where:

- A denotes the distribution of inter-arrival times;
- S denotes the distribution of service times;

---

[1] David George Kendall, English statistician (1918–2007).

– m denotes the number of servers (possibly infinite);

– n denotes the maximum number of customers in the queue (infinite by default).

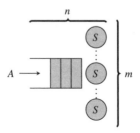

Figure 6.1. *Kendall's notation*

Here A and S typically take their values in the set $\{M, D, E, H, G\}$, where M (for Markovian) refers to the exponential distribution, D the deterministic distribution, E the Erlang distribution (sum of exponentials), H the hyperexponential distribution (random choice among exponentials), and G some general distribution (unspecified). Unless otherwise specified, the customers are served in their order of arrival (First In First Out (FIFO)); we shall see other common service disciplines in section 6.3. For example, the $M/D/1$ denotes a FIFO single-server queue, without any limit on the number of customers, for which customers arrive according to a Poisson process (exponential inter-arrival times, see Chapter 3) and have deterministic service times.

6.2. Traffic and load

We denote the *arrival rate* of customers in the queue by λ, that is denoting by $N(t)$ the number of arrivals in the queue between times 0 and t, the limit:

$$\lambda = \lim_{t \to +\infty} \frac{N(t)}{t} \quad \text{a.s.}$$

The mean inter-arrival time is thus equal to $1/\lambda$.

Similarly, we denote the departure rate of customers from a busy server by μ, or *service rate*; the mean service time is thus equal to $1/\mu$. For an $M/M/\cdot$ queue for instance, customers arrive according to a Poisson process of intensity λ and require a service of exponential duration with parameter μ, as illustrated in Figure 6.2.

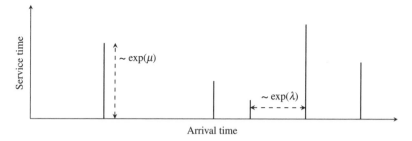

Figure 6.2. *Traffic arriving at an $M/M/\cdot$ queue*

We define the *traffic intensity* as the workload arriving at the queue per time unit. We shall denote it by α. Since customers arrive at rate λ and have service times of mean $1/\mu$, we obtain:

$$\alpha = \frac{\lambda}{\mu}. \qquad [6.1]$$

This is a dimensionless quantity, most often expressed in *erlangs* (symbol E) in the domain of telecommunications (see Chapter 8). When the number m of available servers is infinite, the traffic intensity corresponds to the mean number of busy servers (see the $M/M/\infty$ queue in section 6.4).

We define the *load* as the ratio of the arrival rate to the total service rate of the queue, that is for m servers:

$$\rho = \frac{\lambda}{m\mu}. \qquad [6.2]$$

This is also a dimensionless quantity. Generally, real systems work at load $\rho < 1$: the arrival rate must be less than the total service capacity. As we shall see, this condition ensures the stability of the system for infinite-capacity queues ($n = \infty$) and is necessary to guarantee a low reject probability for finite-capacity queues ($n < \infty$).

6.3. Service discipline

The service discipline defines the order in which customers are served. Common service disciplines are the following:

– FIFO (First In First Out): customers are served in their order of arrival;

– LIFO (Last In First Out): customers are served in their *inverse* order of arrival (in particular, the service of any customer is interrupted by the arrival of a new customer in the queue);

– PS (Processor Sharing[2]): customers are served simultaneously, with fair sharing of the server(s).

While FIFO is the most common service discipline, both in day-to-day queues and buffers of network switches, it is generally not the best in terms of response time: a customer having a long service time forces the other customers to wait, whereas these customers could be served quickly under other service disciplines. For instance, both the LIFO and PS disciplines make it possible for any new customer to start using the server immediately, as illustrated in Figure 6.3. This limits the impact of "heavy" customers and explains why these two service disciplines, and their variants, are implemented to manage the access to the computer processor

---

2 This service discipline was originally conceived for the analysis of multitask, time-shared processor units, hence the name; see the book of Kleinrock quoted in Chapter 1.

units, for instance. We shall see in Chapters 9 and 10 that the PS service discipline is also useful for modeling IP networks, the transfer of information in packets resulting in approximately fair bandwidth sharing between active data flows.

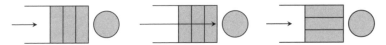

Figure 6.3. *FIFO, LIFO, and PS service disciplines (from left to right)*

6.4. Basic queues

We now present some common queues, all of the $M/M/\cdot$ type. We shall calculate the stationary distribution π of the number of customers $X(t)$ in the queue at time t, including the customer(s) in service. Many performance metrics can be deduced from the stationary distribution. Thus, the mean sojourn time of a customer in the queue, or mean delay δ, follows from the mean number of customers from Little's formula, described in section 6.6:

$$\delta = \frac{E(X)}{\lambda},\qquad\qquad [6.3]$$

where X denotes the number of customers in steady state (the time parameter t is omitted to simplify the notation).

$M/M/1$ *queue*

The simplest and most common queue is the $M/M/1$. Customers arrive according to a Poisson process of intensity λ and require a service of exponential duration with parameter μ. There is a single server and the queue is of infinite capacity.

The number of customers in the queue forms a birth–death process, with birth rate λ and death rate μ. The transition graph of this process is represented in Figure 6.5. In view of equation [6.2], its load is simply given by the ratio of arrival rate to service rate:

$$\rho = \frac{\lambda}{\mu}$$

Figure 6.4. *The M/M/1 queue, a single-server queue*

The stationary measure of the Markov process follows from equation [5.15]:

$$\forall x \in \mathbb{N}, \quad \pi(x) = \pi(0)\rho^x.$$

The queue is stable (in the sense that the number of customers reaches a stationary regime, see section 5.6) if and only if $\rho < 1$. Under this condition, we obtain the stationary distribution by normalization:

$$\forall x \in \mathbb{N}, \quad \pi(x) = (1 - \rho)\rho^x.$$

In particular, the fraction of time the server is busy, $1 - \pi(0)$, is equal to the queue load, ρ (this is in fact a general result, satisfied by the $M/G/1$ queue, see exercise 6 in section 6.11). The mean number of customers in the queue is given by:

$$E(X) = \sum_x x\pi(x) = \frac{\rho}{1 - \rho}.$$

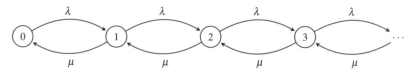

Figure 6.5. *Transition graph of the M/M/1 queue*

From Little's formula [6.3], we deduce the mean delay:

$$\delta = \frac{1}{\mu - \lambda}.$$ [6.4]

$M/M/m$ *queue*

In the presence of m servers, the total service rate depends on the number of customers x: it is equal to $x\mu$ whenever $x \leq m$ and to $m\mu$ otherwise.

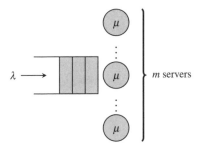

Figure 6.6. *The $M/M/m$ queue, a multiserver queue*

The number of customers in the queue still forms a birth–death process, whose transition graph is shown in Figure 6.7. We deduce the stationary measure:

$$\pi(x) = \pi(0)\frac{\alpha^x}{x!} \qquad \text{if } x \leq m$$

$$\pi(x) = \pi(m)\rho^{x-m} \qquad \text{otherwise.}$$

The queue is stable if and only if $\rho < 1$. Under this condition, we obtain the stationary distribution by normalization:

$$\pi(0) = \frac{1}{1 + \alpha + \frac{\alpha^2}{2} + \ldots + \frac{\alpha^m}{m!} + \frac{\alpha^m}{m!}\frac{\rho}{1-\rho}}.$$

The mean number of customers and the mean delay follow, as for the $M/M/1$ queue.

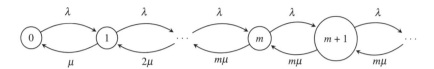

Figure 6.7. *Transition graph of the* $M/M/m$ *queue*

$M/M/\infty$ *queue*

When the number of servers is infinite, there is no waiting: a customer arriving in the system always finds an available server. The total service rate is equal to $x\mu$ in the presence of x customers.

Figure 6.8. *The* $M/M/\infty$ *queue, an infinite-server queue*

The number of customers forms a birth–death process whose transition graph is represented in Figure 6.9 and whose stationary measure is given by:

$$\forall x \in \mathbb{N}, \quad \pi(x) = \pi(0)\frac{\alpha^x}{x!}.$$

The queue is always stable (the load as defined in section 6.2 is null). The stationary distribution is obtained by normalization:

$$\forall x \in \mathbb{N}, \quad \pi(x) = e^{-\alpha}\frac{\alpha^x}{x!}.$$

Thus, the number of customers has a Poisson distribution with mean α in steady state. In particular, the traffic intensity α corresponds to the mean number of busy servers. The mean delay is equal to the mean service time $1/\mu$, which simply follows from the fact that there is no waiting.

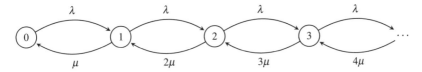

Figure 6.9. *Transition graph of the $M/M/\infty$ queue*

$M/M/m/m$ *queue*

Finally, we consider a queue with m servers that does not accept more than m customers. This is a *loss* system in the sense that a customer who does not find any available server on arrival is blocked and lost; there is no waiting.

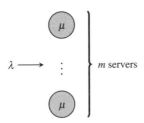

Figure 6.10. *The $M/M/m/m$ queue, a loss multiserver queue*

Again, the number of customers in the queue forms a birth–death process, whose transition graph is depicted in Figure 6.11 (with a virtual transition in state m representing the blocking events) and whose stationary measure is given by:

$$\forall x = 0, 1, \ldots, m, \quad \pi(x) = \pi(0)\frac{\alpha^x}{x!}.$$

The queue is always stable because the state space is finite. The stationary distribution is obtained by normalization:

$$\forall x = 0, 1, \ldots, m, \quad \pi(x) = \frac{\frac{\alpha^x}{x!}}{1 + \alpha + \frac{\alpha^2}{2} + \ldots + \frac{\alpha^m}{m!}}.$$

The probability that all servers are busy is equal to $\pi(m)$. This is also the probability that a customer finds all servers busy on arrival. Applying the formula of conditional transitions of section 5.11, this probability is indeed equal to:

$$\frac{\pi(m)\lambda}{\sum_{x=0}^{m} \pi(m)\lambda} = \pi(m).$$

This is in fact a direct consequence of the PASTA property, which states that customers arriving according to a Poisson process see the system in steady state at the arrival; see section 6.7 below.

We deduce the *blocking rate* of customers, equal to the probability that a customer finds all servers busy on arrival:

$$B = \pi(m) = \frac{\frac{\alpha^m}{m!}}{1 + \alpha + \frac{\alpha^2}{2} + \ldots + \frac{\alpha^m}{m!}}.$$

This is the *Erlang formula*, which will be studied in detail in Chapter 8.

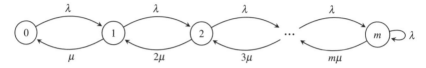

Figure 6.11. *Transition graph of the $M/M/m/m$ queue*

6.5. A general queue

We now consider a general queue of the $M/M/\cdot$ type, with arrival rate λ and total server rate equal to $m(x)\mu$ in state x, as illustrated in Figure 6.12. We can view this system as a single-server queue, where service times have an exponential distribution with parameter μ, the service capacity being equal to $m(x)$ in the presence of x customers. This queue and its extension to finite capacity presented hereafter generalize all basic queues seen so far.

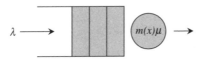

Figure 6.12. *A general queue with state-dependent service rate*

The number of customers in the queue forms a birth–death process, whose transition graph is represented in Figure 6.13 and whose stationary measure is given by:

$$\forall x \in \mathbb{N}, \quad \pi(x) = \pi(0)\frac{\alpha^x}{m(1)\dots m(x)}. \qquad [6.5]$$

Assuming that the sequence $m(x)$ has a (possibly infinite) limit denoted by $\bar{m}$, we redefine the load as:

$$\rho = \frac{\lambda}{\bar{m}\mu}. \qquad [6.6]$$

The queue is stable if and only if $\rho < 1$. The stationary distribution is obtained by normalization.

In view of equation [5.12], the local balance equations are given by:

$$\forall x \in \mathbb{N}, \quad \lambda\pi(x) = m(x+1)\mu\pi(x+1).$$

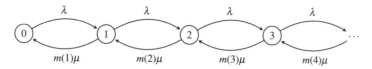

Figure 6.13. *Transition graph of the general queue*

Analyzing them gives the following relationship:

$$\lambda = \sum_{x \geq 1} m(x)\mu\pi(x). \tag{6.7}$$

This is the *conservation law*, which states that in a stable queue the arrival rate of customers must be equal to the departure rate. This law is in fact a direct consequence of the reversibility of the Markov process, the arrivals in one direction of time corresponding to the departures in the other direction of time; see section 5.12.

If the queue has finite capacity n, it is always stable and the stationary distribution is obtained from the normalization of the stationary measure [6.5] over the set of states $\{0, 1, \ldots, n\}$. Summing the local balance equations:

$$\forall x = 0, 1, \ldots, n - 1, \quad \lambda\pi(x) = m(x + 1)\mu\pi(x + 1),$$

we obtain:

$$\lambda = \sum_{x=1}^{n} m(x)\mu\pi(x) + \lambda\pi(n).$$

This is a new form of the conservation law, which accounts for the fact that in state $x = n$ the departure rate of customers is equal to the sum of the departure rate of admitted customers, $m(n)\mu$, and of rejected customers, λ. Since the blocking rate is given by $B = \pi(n)$ (using the same argument as for the $M/M/m/m$ queue), this law can also be written as:

$$\lambda(1 - B) = \sum_{x=1}^{n} m(x)\mu\pi(x). \tag{6.8}$$

The law states that, in a finite-capacity queue, the arrival rate of *admitted* customers is equal to the departure rate of *served* customers. Again, this is in fact a direct consequence of the reversibility of the associated Markov process.

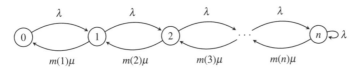

Figure 6.14. *Transition graph of the general finite-capacity queue*

6.6. Little's formula

Consider a queue in steady state. If the arrival rate of customers in this queue is equal to λ, we have the following simple relationship between the mean number of customers in the queue and the mean time spent by each customer in the queue, δ:

$$E(X) = \lambda\delta. \tag{6.9}$$

This is Little's formula, valid for any queue, and more generally any system receiving a stream of customers, in steady state. In particular, the result applies to the queuing networks described in Chapter 7.

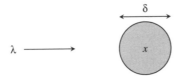

Figure 6.15. *Little's formula:* $E(X) = \lambda\delta$

To prove this relationship, it is sufficient to calculate the cumulated number of customers in the queue between times 0 and T, that is:

$$\bar{X}(T) = \int_0^T X(t)\,\mathrm{d}t.$$

From the ergodic theorem [5.7], we have:

$$\lim_{T \to \infty} \frac{\bar{X}(T)}{T} = \mathrm{E}(X). \qquad\qquad [6.10]$$

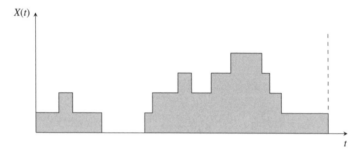

Figure 6.16. *Evolution of the number of customers in the queue*

Moreover, denoting by $N(t)$ the number of customers who have arrived in the queue between times 0 and t, and $\delta_1, \delta_2, \dots$ the sequence of time spent by successive customers in the queue, we have:

$$\bar{X}(T) = \sum_{n=1}^{N(T)} \delta_n + o(T).$$

This relationship follows from the fact that, up to the border effects, $\bar{X}(T)$ is the area of the surface formed by the set of rectangles associated with the sojourns of customers in the queue, each rectangle having width 1 and length equal to the sojourn time of the customer in the queue. We obtain:

$$\lim_{T \to \infty} \frac{\bar{X}(T)}{T} = \lim_{T \to \infty} \frac{1}{T} \sum_{n=1}^{N(T)} \delta_n,$$

$$= \lim_{T \to \infty} \frac{N(T)}{T} \frac{1}{N(T)} \sum_{n=1}^{N(T)} \delta_n$$

$$= \lambda \delta. \hspace{3cm} [6.11]$$

Little's formula is deduced from equations [6.10] and [6.11].

6.7. PASTA property

When arriving according to a Poisson process, customers "see" the system in steady state at their arrival. Thus, the probability that a customer sees the queue in state x on arrival in the queue is equal to $\pi(x)$. From the ergodic theorem, $\pi(x)$ is also the fraction of time the system spends in state x. We call this property PASTA for "Poisson Arrivals See Time Averages". It is a key property of Poisson processes, due to the fact that arrivals are mutually independent, also knowing the arrival time of one customer does not give any information on the arrival times of the other customers (see Chapter 3), and in particular on the arrival times of the customers present in the queue. The other point processes do not have this property, as shown by exercise 3 in section 6.11.

6.8. Insensitivity

While it is convenient to assume that the service time distribution is exponential, this assumption is rarely satisfied in practice. Fortunately, some service disciplines have the so-called insensitivity property, for which the stationary distribution of the state of the queue is independent of the distribution of service times beyond the mean. This is the case of the LIFO and PS service disciplines, for instance (see section 6.3). Thus, under these service disciplines, the stationary distributions derived in section 6.4 for basic $M/M/\cdot$ queues, and more generally in section 6.5 for an $M/M/\cdot$ queue with state-dependent service rates, are identical to those of the equivalent $M/G/\cdot$ queues with the same mean service time. Exercise 13 in section 6.11 provides a proof of the insensitivity property of the PS service discipline.

The FIFO service discipline is *not* insensitive. This is due to the presence of heavy customers, which increase the mean delay in the queue. The Pollaczek–Khinchin formula presented in the next section enables us to precisely quantify this phenomenon for an $M/G/1$ queue. Specifically, the mean delay increases linearly with the *variance* of the service times.

REMARK 6.1.– (Infinite-server queue). The service discipline is irrelevant for infinite-server queues, since there is no waiting. In particular, the $M/G/\infty$ queue is always insensitive (see exercise 14 in section 6.11).

6.9. Pollaczek–Khinchin's formula

Consider an $M/G/1$ queue of load $\rho < 1$ under the FIFO service discipline. Since the distribution of the number of customers in the queue is generally unknown, we seek to calculate directly the mean sojourn time δ. We assume that the service time, represented by the random variable S, has a finite variance.

An arriving customer must wait till the customer in service, if any, leaves the queue, which takes τ time units on average, and till the other waiting customers, if any, are served. To calculate τ, we need the mean service time of the customer in service, if any. This time is *not* equal, in general, to the mean service time $E(S)$, but to the ratio[3] $E(S^2)/E(S)$. We deduce the mean time necessary for the customer in service to leave the queue:

$$\tau = \frac{E(S^2)}{2E(S)}. \tag{6.12}$$

The probability that there is a customer in service in the queue is, in view of the PASTA property, the probability that the queue is not empty, which is equal to ρ (see exercise 6 in section 6.11). Thus, it takes on average $\rho\tau$ time units for the server to be available for a new customer.

It remains to calculate the time it takes to serve the waiting customers. Let $X' = \max(X - 1, 0)$ be the number of waiting customers in steady state. From the PASTA property, this is also the distribution of the number of waiting customers at the arrival of the considered customer. Each of these customers requires a mean service time equal to $E(S)$. We deduce the mean waiting time of the considered customer:

$$\delta' = \rho\tau + E(X')E(S).$$

In view of Little's formula, we have:

$$E(X') = \lambda\delta'.$$

Since $\rho = \lambda/\mu$, we deduce:

$$\delta' = \rho\tau + \rho\delta',$$

that is:

$$\delta' = \frac{\rho}{1-\rho}\tau.$$

---

3 This is related to the famous "bus paradox" (see exercise 5 in section 3.9): just as a customer is more likely to arrive at a bus stop during a long time interval between two buses, a customer is more likely to arrive in the queue during the service of a customer that requires a long service time.

Adding the mean service time and using equation [6.12], we obtain the mean sojourn time in an $M/G/1$ queue, with $1/\mu = \mathrm{E}(S)$:

$$\delta = \frac{1}{\mu}\left(1 + \frac{\rho}{1 - \rho}\frac{\mathrm{E}(S^2)}{\mathrm{E}(S)^2}\right).$$

This is *Pollaczek–Khinchin's formula*[4]. In particular, we find the mean delay in an $M/M/1$ queue, equation [6.4], the ratio $\mathrm{E}(S^2)/\mathrm{E}(S)^2$ being equal to 2 in this case. We observe that the mean sojourn time in an $M/G/1$ queue under the FIFO service discipline is very sensitive to the service time distribution: it increases linearly with the variance of the service times.

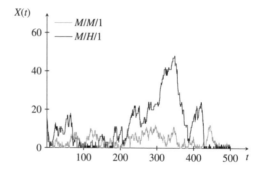

Figure 6.17. *Evolution of the number of customers in $M/M/1$ and $M/H/1$ queues*

This sensitivity is illustrated in Figure 6.17, which compares the evolution of the number of customers in $M/M/1$ and $M/H/1$ queues of same load $\rho = 0.8$, with unit mean service times. The hyperexponential distribution is the mix of an exponential distribution with parameter $\mu_1 = 1/4$, with probability $p_1 = 1/5$, and an exponential distribution with

4 Félix Pollaczek, French-Austrian mathematician (1892–1981); Alexander Khinchin, Russian mathematician (1894–1959).

parameter $\mu_2 = 4$, with probability $p_2 = 4/5$; the mean service time, given by $p_1/\mu_1 + p_2/\mu_2$, is equal to 1. Pollaczek–Khinchin's formula gives $\delta = 5$ for the $M/M/1$ queue and $\delta \approx 14$ for the $M/H/1$ queue: the presence of "heavy" customers has a strong impact on the mean sojourn time of all customers.

6.10. The observer paradox

While the PASTA property gives the state experienced by customers at their arrival, provided the arrival process is Poisson, it remains to calculate the state experienced by a customer present in the queue. We shall see that conditioning on the presence of a customer changes the distribution of the system state. This is the *observer paradox*. We restrict the analysis to the PS service discipline.

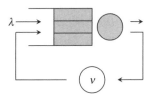

Figure 6.18. *A queue with a "tagged" customer*

We consider a general queue, as defined in section 6.5, with infinite capacity. In addition to "standard" customers, who arrive according to a Poisson process of intensity λ, a "tagged" customer periodically visits the queue, as illustrated in Figure 6.18. When present in the queue, this customer cannot be distinguished from the other customers; in particular, the customer requires a service of exponential duration with parameter μ and must share the service capacity evenly with the other customers. After service completion, the tagged customer remains idle during an exponential duration with parameter ν before re-entering the queue.

The tagged customer will be used to study the state experienced by an arbitrary customer in the queue. We denote by $X(t)$ the number of standard customers in the queue at time t, and by $Y(t)$ the state of the tagged customer at time t: $Y(t) = 1$ if the customer is in the queue, $Y(t) = 0$ otherwise. The evolution of the system state $(X(t), Y(t))$ defines a Markov process on $\mathbb{N} \times \{0, 1\}$, whose transition graph is shown in Figure 6.19. This process is reversible, with stationary measure:

$$\pi(x, 0) = \pi(0, 0)\frac{\alpha^x}{m(1)\dots m(x)},$$

$$\pi(x, 1) = \pi(0, 0)(x + 1)\frac{\alpha^x \beta}{m(1)\dots m(x + 1)},$$

[6.13]

with $\beta = \nu/\mu$. The stability condition is the same as in the absence of the tagged customer, namely $\rho < 1$, where ρ denotes the queue load, given by equation [6.6]. The stationary distribution is obtained by normalization.

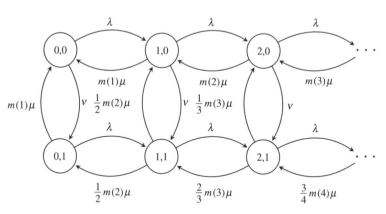

Figure 6.19. *Transition graph of the Markov process describing the system state*

Denote by π' the stationary distribution of the total number of customers seen by the tagged customer, including self, when present in the queue. The probability $\pi'(x)$ that the total number of customers is equal to x is proportional to $\pi(x-1,1)$. In view of equations [6.5] and [6.13], we obtain:

$$\pi'(x) \propto x\pi(x),$$

where π denotes the stationary distribution of the initial system, without the tagged customer. We have indeed a bias, given by the stationary distribution $\pi(x)$ weighted by the number of customers x. Thus, the tagged customer is more likely to see a large number of customers in the queue, which is a simple consequence of the fact that the tagged customer consumes part of the service capacity. After normalization, we obtain:

$$\pi'(x) = \frac{x\pi(x)}{\mathrm{E}(X)}. \qquad [6.14]$$

It remains to verify that the tagged customer sees the initial system in steady state on arrival so that we have indeed observed the system from the viewpoint of an arbitrary customer (there is no bias related to the tagged customer). The PASTA property is not applied since the tagged customer does not arrive according to a Poisson process. However, we know from the formula of conditional transitions (see section 5.11) that the probability that the tagged customer finds x standard customers in the queue on arrival is proportional to $\pi(x,0)$. From equations [6.5] and [6.13], this probability is equal to $\pi(x)$, which corresponds to the stationary distribution of the initial system, without tagged customer.

The argument is the same for a finite-capacity queue. If the queue if full on arrival, the tagged customer returns immediately to idle state. The associated Markov process $(X(t), Y(t))$ is the restriction of the previous process to those states (x, y) such that $x + y \leq n$, where n is the queue capacity. The stationary distribution π' of the total number of customers seen by the tagged customer is again given by the initial distribution π, weighted by the number of customers.

6.11. Exercises

1. *Departures from an $M/M/1$ queue*
Show that the departure process of customers from an $M/M/1$ queue at equilibrium is a Poisson process. Moreover, show that the number of customers in the queue at time t is independent of the departure process before time t. We might use the reversibility of the associated Markov process.

2. *PASTA*
Consider a general $M/M/\cdot$ queue as described in section 6.5, with infinite capacity and stable. Using the formula of conditional transitions described in section 5.11, give the stationary distribution of the state of the queue seen by customers at their arrival in the queue. Verify the PASTA property.

3. *Anti-PASTA*
Consider a $D/D/1$ queue. Customers arrive in the queue every two time units, each requiring a service of one time unit. Compare the stationary distributions of the number of customers at an arbitrary time and at the arrival time of a customer.

4. *Arrival rate*
Prove, using Little's formula, that the arrival rate of customers is equal to the inverse of the mean inter-arrival time.

5. *Admission rate*
Calculate using the technique described in section 5.8 the rate (per time unit) of admitted customers in an $M/M/m/m$

queue. Express this quantity as a function of the customer arrival rate and the blocking probability. Find the latter expression directly from Little's formula.

6. Occupancy rate
Show that the occupancy rate (i.e. the probability that the queue is not empty) of a stable $M/G/1$ queue is equal to its load, ρ. We might apply Little's formula on the server, assuming the service discipline is FIFO, then notice that the result is independent of the service discipline.

7. Delay distribution
Show that the delay in a FIFO $M/M/1$ queue has an exponential law with parameter $\mu - \lambda$.

8. Little's formula
Using the technique described in section 6.10, prove Little's formula for the general queue introduced in section 6.5 under the PS service discipline.

9. Finite-capacity queue
Calculate the reject probability of a customer in an $M/M/1/n$ queue, as well as the mean sojourn time of admitted customers.

10. Infinite-server queue
Show that the mean number of customers in an $M/G/\infty$ queue is equal to the traffic intensity.

11. Service discipline
Show using the property of the exponential distribution described in section 2.4 that at any time where an $M/M/1$ queue is not empty, the time elapsed before the next departure of a customer has an exponential distribution with parameter μ, whatever the service discipline. Deduce that the stationary distribution of the state of an $M/M/1$ queue is independent of the service discipline.

12. *Sensitivity of FIFO queues*

Using Pollaczek–Khinchin's formula calculate the mean sojourn time in $M/D/1$, $M/M/1$, and $M/H/1$ FIFO queues. For the latter queue, the service times are exponential with parameter μ_1 with probability p_1 and exponential with parameter μ_2 with probability p_2, with $p_1 + p_2 = 1$.

13. *Insensitivity of PS queues*

Consider an $M/H/1$ queue under the PS service discipline. Customers arrive according to a Poisson process of intensity λ; the service times are exponential with parameter μ_1 with probability p_1 and exponential with parameter μ_2 with probability p_2, with $p_1 + p_2 = 1$. The mean service time is thus equal to $1/\mu = p_1/\mu_1 + p_2/\mu_2$, and the load to $\rho = \lambda/\mu$. We assume $\rho < 1$. Give the stationary distribution of the state of the queue, distinguishing customers according to the required service type; show that the total number of customers has the same distribution as that in an $M/M/1$ queue of load ρ.

14. *Insensitivity of infinite-server queues*

Consider an $M/G/\infty$ queue with arrival rate λ and service rate μ. Using exercise 4 in section 3.9 and assuming that the service times take a discrete set of values, show that the number of customers in the queue has a Poisson distribution with mean $\alpha = \lambda/\mu$.

15. *Printers*

A small company is equipped with a single printer, which does not suffice its needs. It has the following options:

1) replace this printer by a more recent printer, twice as fast;

2) buy another printer of the same type and balance the load on the two printers (each request is forwarded to each printer with the same probability);

3) buy another printer of the same type and use a server to queue the waiting requests and forward them to the first available printer.

Compare the mean response times of these three systems, assuming requests arrive according to a Poisson process and have exponential service times.

6.12. Solution to the exercises

1. Departures from an $M/M/1$ queue

The process $X(t)$ describing the number of customers in the queue is reversible (see section 5.13), thus $X(t)$ has the same distribution as the process in reverse time, $\tilde{X}(t)$. In particular, the times of positive jumps of $X(t)$ and $\tilde{X}(t)$ have the same distributions: both define a Poisson process of intensity λ. But the times of positive jumps of $\tilde{X}(t)$ are the times of negative jumps of $X(t)$, that is the departure process from the queue.

Moreover, it is obtained from the memoryless property of Poisson processes that the state of the queue at time t is independent of the arrivals before time t. Applying this argument to the process in reverse time, we deduce that the state of the queue at time t is independent of the departures before time t.

2. PASTA

Let π be the stationary distribution of the number of customers in the queue. Applying the formula of conditional transitions, an incoming customer sees the queue in state x with probability:

$$\frac{\pi(x)\lambda}{\sum_{y \geq 0} \pi(y)\lambda} = \pi(x).$$

The PASTA property is therefore verified.

3. Anti-PASTA

The number of customers in the queue defines a periodic process $X(t)$, as depicted by the following graph, assuming the first customer arrives at time $t = 0$:

The stationary distribution π of the state of the queue at an arbitrary time is given by $\pi(0) = 1/2$ and $\pi(1) = 1/2$. When a customer arrives in the queue, the queue is always empty. The PASTA property is thus not verified.

4. Arrival rate

Consider a queue such that each customer is served until the arrival of the next customer. The number of customers in the queue is constant, equal to 1, and the mean service time of a customer is equal to the mean inter-arrival time. Using Little's formula, we deduce that the mean inter-arrival time is equal to the inverse of the arrival rate of customers (see section 6.2).

5. Admission rate

Denote by λ' the rate of admitted customers. Considering the corresponding transitions of the Markov process, we get:

$$\lambda' = \lambda \sum_{x=0}^{m-1} \pi(x)$$

where π is the stationary distribution of the number of customers in the queue. We deduce $\lambda' = \lambda(1 - B)$, where $B = \pi(m)$ is the blocking rate.

The mean service time of a customer is equal to $1/\mu$ if the customer is admitted and is null otherwise; the mean sojourn time is thus equal to $(1 - B)/\mu$. Using Little's formula, we have $E(X) = \lambda(1 - B)/\mu$. Moreover, the mean service time of admitted customers is equal to $1/\mu$. Again applying Little's formula, we deduce the rate of admitted customers: $\lambda' = \lambda(1 - B)$.

6. Occupancy rate

We first consider the FIFO service discipline. Since the queue is stable, the arrival rate at the server is equal to the arrival rate at the queue, that is λ. The mean delay of a customer in the server is equal to the mean service time, $1/\mu$. By Little's formula, we obtain the mean number of customers in the server: $\rho = \lambda/\mu$. We conclude by observing that the mean number of customers in the server is equal to the probability that the server is busy, that is the occupancy rate. Since the work provided by the server per time unit is independent of the order in which customers are served, the occupancy rate is the same for all service disciplines, equal to ρ.

7. Delay distribution

A customer who arrives in a FIFO queue must wait till all the customers arrived before them are served. From the PASTA property, this customer sees the queue in a steady state. The number of customers in the queue who arrived before the considered customer has a geometric distribution on $\mathbb{N}$ with parameter $1 - \rho$. In view of section 2.6, the waiting time has an exponential distribution with parameter $\mu(1-\rho) = \mu - \lambda$.

8. Little's formula

Consider the process defined in section 6.10, with stationary distribution π. Denote by δ the mean sojourn time in the queue. This is also the mean sojourn time of the observer, who then visits the queue every $\delta + 1/\nu$ time units on average. But the frequency of visits to the queue by the observer is given by:

$$\sum_{x=0}^{\infty} \pi(x, 0)\nu.$$

From equations [6.5] and [6.13], we deduce:

$$\sum_{x=0}^{\infty} \pi(x, 0) = \frac{\pi(0, 0)}{\pi(0)} \quad \text{and} \quad \sum_{x=0}^{\infty} \pi(x, 1) = \frac{\pi(0, 0)}{\pi(0)} E(X) \frac{\nu}{\lambda},$$

where $\pi(0)$ denotes the probability that the queue is empty. Thus, we have:

$$\sum_{x=0}^{\infty} \pi(x, 0) = \frac{1}{1 + E(X)\frac{\nu}{\lambda}}.$$

Writing the equality of frequencies of visits of the observer:

$$\frac{\nu}{1 + E(X)\frac{\nu}{\lambda}} = \frac{1}{\delta + \frac{1}{\nu}},$$

we obtain Little's formula:

$$\lambda\delta = E(X).$$

9. *Finite-capacity queue*

The Markov process describing the number of customers in an $M/M/1$ queue being reversible, the results of section 5.15 apply and the stationary distribution of the $M/M/1/n$ queue is obtained as:

$$\pi(x) = \frac{\rho^x}{1 + \rho + \ldots + \rho^n}, \quad \forall x = 0, \ldots, n.$$

We deduce the mean number of customers in the queue:

$$E(X) = \begin{cases} \dfrac{n\rho^{n+2} - (n+1)\rho^{n+1} + \rho}{(1 - \rho^{n+1})(1 - \rho)} & \text{if } \rho \neq 1 \\ \dfrac{n}{2} & \text{if } \rho = 1. \end{cases}$$

Moreover, the arrival rate of admitted customers is given by $\lambda(1 - \pi(n))$ (see exercise 5). Applying Little's formula, we deduce the mean sojourn time of a customer admitted in the queue:

$$\delta = \begin{cases} \dfrac{n\rho^{n+1} - (n+1)\rho^n + 1}{\mu(1 - \rho^{n+1})(1 - \rho)} & \text{if } \rho \neq 1 \\ \dfrac{n}{2\mu} & \text{if } \rho = 1. \end{cases}$$

10. Infinite-server queue

In an $M/G/\infty$ queue, each customer is served immediately. The mean sojourn time is thus equal to the mean service time, $1/\mu$. Applying Little's formula, we obtain that the mean number of customers is equal to the traffic intensity $\alpha = \lambda/\mu$.

11. Service discipline

Let p_k be the fraction of service rate devoted to the kth customer (this depends on the service discipline). This customer virtually leaves the queue after an exponential duration with parameter $p_k\mu$. Thus, the next departure occurs after an exponential duration with parameter $\mu \sum_{k \geq 1} p_k = \mu$ (see section 2.4). We deduce that the stationary distribution of the state of an $M/M/1$ queue is independent of the service discipline.

12. Sensitivity of FIFO queues

Applying Pollaczek–Khinchin's formula, we obtain for $M/D/1$, $M/M/1$, and $M/H/1$ queues, respectively:

$$\delta_D = \frac{1 - \rho/2}{\mu - \lambda}, \quad \delta_M = \frac{1}{\mu - \lambda}, \quad \delta_H = \frac{1}{\mu} + \left(\frac{p_1}{\mu_1^2} + \frac{p_2}{\mu_2^2}\right)\frac{\rho}{\mu - \lambda}.$$

The mean sojourn time depends on the service time distribution beyond the mean: the FIFO service discipline is sensitive.

13. Insensitivity of PS queues

The number of customers of each type defines a Markov process $(X_1(t), X_2(t))$ whose transition graph is shown in Figure 6.20.

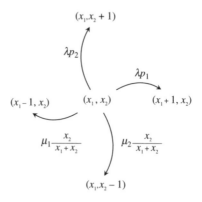

Figure 6.20. *Transition rates from state x for an $M/H/1$ queue under the PS service discipline*

This is a reversible process, with stationary measure:

$$\pi(x_1, x_2) = \pi(0,0) \binom{x_1 + x_2}{x_1} \rho_1^{x_1} \rho_2^{x_2},$$

with $\rho_1 = \lambda p_1/\mu_1$ and $\rho_2 = \lambda p_2/\mu_2$. Since $\rho = \rho_1 + \rho_2$, we deduce that the queue is stable if and only if $\rho < 1$. The stationary distribution is given by:

$$\pi(x_1, x_2) = (1 - \rho) \binom{x_1 + x_2}{x_1} \rho_1^{x_1} \rho_2^{x_2}.$$

The stationary distribution of the total number of customers obtained follows by summation:

$$\pi(x) = \sum_{x_1, x_2 : x_1 + x_2 = x} \pi(x_1, x_2) = (1 - \rho)\rho^x.$$

This is the stationary distribution of the number of customers in an $M/M/1$ queue.

NOTE: The result extends to any phase-type distribution, namely mixtures of Erlang distributions, which are known to be dense within the set of all distributions of positive real random variables. This shows the insensitivity property of the PS service discipline.

14. Insensitivity of infinite-server queues

Assume that the service time of a customer is equal to t_k with probability p_k, with $\sum_k p_k t_k = 1/\mu$. From the property of subdivision of Poisson processes (see section 3.6), those customers who have service time t_k arrive according to a Poisson process of intensity λp_k. In view of exercise 4 in section 3.9, their number in the queue has a Poisson distribution with mean $\lambda p_k t_k$. Thus, the total number of customers in the queue has a Poisson distribution with mean $\lambda \sum_k p_k t_k = \lambda/\mu$.

15. Printers

Let λ be the request arrival rate and μ be the service rate of the original printer. Solution 1 corresponds to an $M/M/1$ queue with arrival rate λ and service rate 2μ. Under the stability condition $\lambda < 2\mu$, the mean response time is given by:

$$\delta_1 = \frac{1}{2\mu - \lambda}$$

Solution 2 corresponds to two $M/M/1$ queues in parallel, each with arrival rate $\lambda/2$ (from the property of subdivision of a Poisson process) and service rate μ. Under the stability condition $\lambda < 2\mu$, the mean response time is given by:

$$\delta_2 = \frac{2}{2\mu - \lambda}$$

Solution 3 corresponds to an $M/M/2$ queue with arrival rate λ and service rate μ per server. Under the stability condition

$\lambda < 2\mu$, the mean response time is given by:

$$\delta_3 = \frac{1}{\mu(1 - \rho^2)},$$

with $\rho = \lambda/2\mu$. Hence, $\delta_1 < \delta_3 < \delta_2$ is verified.

Chapter 7

Queuing Networks

This chapter discusses networks of queues, useful for modeling systems with multiple service units.

7.1. Jackson networks

We first consider a network consisting of N single-server queues. Customers arrive at queue i according to a Poisson process of intensity ν_i. We denote by $\nu = \sum_{i=1}^{N} \nu_i$ the total customer arrival rate, which we assume to be positive. Equivalently, customers arrive in the network according to a Poisson process of intensity ν and join queue i with probability ν_i/ν (see section 3.6).

After service completion at queue i, a customer moves to queue j with probability p_{ij} and leaves the network with probability:

$$p_i = 1 - \sum_{j=1}^{N} p_{ij}.$$

Service times at queue i have an exponential distribution with parameter μ_i. Such a network is called a *Jackson network*; an example of this is given in Figure 7.1.

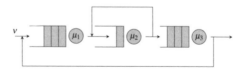

Figure 7.1. *A Jackson network of $N = 3$ queues*

Let $X_i(t)$ be the number of customers in the queue i at time t. We refer to the associated vector $X(t)$ as the *network state*. This is a Markov process on $\mathbb{N}^N$, whose transition rates are represented in Figure 7.2.

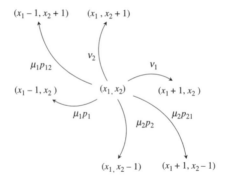

Figure 7.2. *Transition rates from state x for a network of $N = 2$ queues*

7.2. Traffic equations

The stationary distribution of the network state depends on the total arrival rate at each queue. Intuitively, denoting by λ_i the arrival rate of customers at queue i, we must have:

$$\lambda_i = \nu_i + \sum_{j=1}^{N} \lambda_j p_{ji}, \quad i = 1, \ldots, N, \tag{7.1}$$

the second term corresponding to *internal* arrivals, due to customers moving from one queue to another. These are the *traffic equations*, which have a unique solution, as shown below. Note that, by summing these equations, we obtain the *conservation law* of the network, which states that the total arrival rate in the network must be equal to the total departure rate from the network:

$$\nu = \sum_{i=1}^{N} \lambda_i p_i. \tag{7.2}$$

To obtain the traffic equations, we describe the random path followed by a customer in the network. This path is described by a Markov chain Z_n, $n \in \mathbb{N}$, on $\{0, 1, \ldots, N\}$; state 0 corresponds both to the *source* from which the customer arrives at her arrival in the network, and to the *sink* where the customer joins when leaving the network; state i corresponds to the presence of the customer at queue i. The transition probabilities of the Markov chain are thus given by the routing probabilities p_{ij}, for all $i, j \in \{1, \ldots, N\}$, $p_{00} = 0$, $p_{0i} = \nu_i/\nu$, and $p_{i0} = p_i$ for all $i \in \{1, \ldots, N\}$. Figure 7.3 shows, for instance, the transition graph of the Markov chain associated with the network of Figure 7.1.

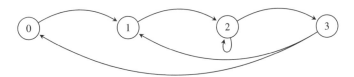

Figure 7.3. *Transition graph of the Markov chain associated with the routing of a customer in the network of Figure 7.1*

In the following sections, we assume that the Markov chain Z_n is *irreducible*, which means that each customer eventually leaves the network. The chain Z_n has a finite state space and thus a unique stationary distribution $\pi_0, \pi_1, \ldots, \pi_N$ solution to the balance equations:

$$\pi_0 = \sum_{i=1}^{N} \pi_i p_i \quad \text{and} \quad \pi_i = \pi_0 \frac{\nu_i}{\nu} + \sum_{j=1}^{N} \pi_j p_{ji}, \quad i = 1, \ldots, N, \quad [7.3]$$

to which we add the normalization condition:

$$\sum_{i=0}^{N} \pi_i = 1. \qquad [7.4]$$

Since π_i is the frequency of visit to state i, for all $i = 0, \ldots, N$, and $1/\pi_0$ the mean return time to state 0 (see section 4.8), the ratio π_i/π_0 corresponds to the mean number of visits of queue i by any customer, between their arrival in the network and their departure from the network. Since the customers arrive from outside at rate ν, the arrival rate of the customers at queue i is given by $\lambda_i = \nu \pi_i/\pi_0$. We verify from equation [7.3] that arrival rates are indeed defined as the unique solution to the traffic equations [7.1], the normalizing condition [7.4] corresponding to the conservation law [7.2].

7.3. Stationary distribution

It turns out that, under the stability condition, the stationary distribution of the Markov process $X(t)$ is given by the product of the stationary distributions of the number of customers in each queue, that is:

$$\pi(x) = \prod_{i=1}^{N} (1 - \rho_i) \rho_i^{x_i}, \quad x \in \mathbb{N}^N, \qquad [7.5]$$

where $\rho_i = \lambda_i/\mu_i$ denotes the load of queue i. We say that the stationary distribution has a *product form*. This is a key result of the queuing theory, shown by Jackson in 1963.

Note that, at equilibrium, the number of customers in each queue is independent of the number of customers in other queues. Everything happens as if customers arrived in queue i according to a Poisson process of intensity λ_i, for all $i = 1, \ldots, N$. In particular, the mean sojourn time in queue i is given by that in the corresponding $M/M/1$ queue, that is:

$$\delta_i = \frac{1}{\mu_i - \lambda_i}.$$

To show Jackson's result, it is sufficient to verify that the stationary measure of the Markov process $X(t)$ is:

$$\pi(x) = \pi(0) \prod_{i=1}^{N} \rho_i^{x_i}, \quad x \in \mathbb{N}^N. \tag{7.6}$$

We indeed deduce from this expression the stability condition, that is $\rho_i < 1$ for each queue i, and, under this condition, the stationary distribution [7.5] after normalization. To show that equation [7.6] gives the stationary measure of the Markov process $X(t)$, we can simply verify the corresponding balance equations (see section 5.4). In fact, it is sufficient to verify the following *partial* balance equations, for all $i = 1, \ldots, N$:

$$\pi(x)\mu_i 1(x_i \geq 1) = \pi(x - e_i)\nu_i + \sum_{j=1}^{N} \pi(x - e_i + e_j)\mu_j p_{ji}, \tag{7.7}$$

$$\pi(x)\nu = \sum_{i=1}^{N} \pi(x + e_i)\mu_i p_i, \tag{7.8}$$

where e_i denotes the unit vector on component i, and with the convention that $\pi(x) = 0$ if $x \notin \mathbb{N}^N$. Replacing π by its expression [7.6], we find the traffic equations [7.1] and [7.2]. Equation [7.7] states that, at equilibrium, the frequency of departures from queue i in state x must be equal to that of arrivals in this queue *leading* to state x; equation [7.8], which follows from the N others, states that the frequency of arrivals in the network in state x must be equal to the frequency of departures from the network *leading* to state x. These equations are stronger than the balance equations of the Markov process $X(t)$ since the latter can be obtained by summation:

$$\pi(x) \sum_{i=1}^{N} (\nu_i + \mu_i 1(x_i \geq 1)) = \sum_{i=1}^{N} \pi(x - e_i)\nu_i$$

$$+ \sum_{i,j=1}^{N} \pi(x - e_i + e_j)\mu_j p_{ji} \quad [7.9]$$

$$+ \sum_{i=1}^{N} \pi(x + e_i)\mu_i p_i.$$

Note that the balance equations [7.9] account for the potential virtual jumps of the Markov process $X(t)$ (see section 5.9), when a customer immediately re-enters the queue the customer leaves (which may occur at queue i if and only if $p_{ii} > 0$).

7.4. MUSTA property

Like simple queues, one can be interested in the state of the network seen by a customer on arrival in a given queue. For external arrivals, the "Poisson Arrivals See Time Averages" (PASTA) property of Poisson processes applies: the customer sees, just before the arrival, the network in its stationary state, according to distribution [7.5]. It turns out that the same property holds for internal arrivals, whereas these do not form in general a Poisson process. This is the MUSTA property for Moving Units See Time Averages.

To show this property, we calculate the frequency of the associated events (see section 5.11). The frequency of the event "a customer moves from queue i to queue j and finds the network in state x just before her arrival at queue j" is equal to $\pi(x + e_i)\mu_i p_{ij}$, that is, in view of equation [7.6]:

$$\pi(x)\lambda_i p_{ij}.$$

The frequency of moves from queue i to queue j being equal to $\lambda_i p_{ij}$, the probability that a customer moving from queue i to queue j finds the network in state x just before her arrival is equal to $\pi(x)$: this is the probability that the network is in state x at equilibrium.

Similarly, a customer leaving the network from state i sees the network in steady state after their departure. The proof is similar.

7.5. Closed networks

The networks considered so far are *open* in the sense that customers arrive and eventually leave the network after having followed a random path in the network. Here we consider a *closed* network for which there is a fixed number of customers, denoted by M. There are neither arrivals in the network nor departures from the network. With the notations of section 7.1, we have $\nu = 0$ (no external arrivals) and $p_i = 0$ for all $i = 1, \ldots, N$ (no departures from the network). Figure 7.4 shows an example of a closed network.

We assume that each customer can visit all the queues; denoting by λ_i the relative frequency of visit of a customer to queue i, we must have:

$$\lambda_i = \sum_{j=1}^{N} \lambda_j p_{ji}, \quad i = 1, \ldots, N. \tag{7.10}$$

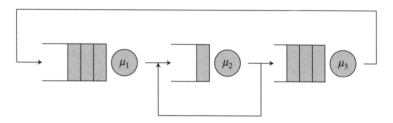

Figure 7.4. *A closed network of N = 3 queues*

These are the *traffic equations*. Like for open networks, we show them by considering the sequence of queues visited by an arbitrary customer. This random path defines a Markov chain Z_n, $n \in \mathbb{N}$, on $\{1, \ldots, N\}$. This chain is irreducible on a finite state space and thus has a unique stationary distribution. The traffic equations [7.10] are the corresponding balance equations.

We then define the relative load of queue i as the ratio $\rho_i = \lambda_i / \mu_i$. A stationary measure of the Markov process $X(t)$ describing the evolution of the network is given by:

$$\pi(x) = \prod_{i=1}^{N} \rho_i^{x_i}, \quad \forall x \in \mathbb{N}^N : \sum_{i=1}^{N} x_i = M. \qquad [7.11]$$

Since the state space is finite, the network is always stable and the stationary distribution follows by normalization. We deduce the mean delays from Little's formula by first calculating the arrival rates of customers at each queue (see exercise 6 in section 7.8).

To show that the stationary measure is indeed given by equation [7.11], we can, as for open networks, verify the partial balance equations:

$$\pi(x)\mu_i 1(x_i \geq 1) = \sum_{j=1}^{N} \pi(x - e_i + e_j)\mu_j p_{ji}, \quad i = 1, \ldots, N. \quad [7.12]$$

Replacing π by its expression [7.11], we find the traffic equations [7.10]. Equations [7.12] state that, at equilibrium, the frequency of departures from queue i in state x must be equal to that of arrivals in this queue *leading* to state x. The balance equations of the Markov process $X(t)$ follow by summation.

7.6. Whittle networks

In Whittle networks, the total service rate of each queue depends on the network state. Customers require exponential service times with parameter μ_i at queue i, the service capacity being equal to $m_i(x)$ in state x; the total service rate of queue i is thus equal to $\mu_i m_i(x)$ in state x.

We start by assuming that customers arrive according to a Poisson process of intensity λ_i at queue i, then leave the network after service completion (there is no routing). The network is said to be a Whittle network if the service capacities satisfy the following balance equation:

$$m_i(x)m_j(x - e_i) \hspace{4cm} [7.13]$$
$$= m_j(x)m_i(x - e_j), \quad \forall i, j, \ \forall x : x_i > 0, x_j > 0.$$

We deduce from Kolmogorov's criterion (see section 5.14) that this condition is equivalent to the reversibility of the Markov process $X(t)$ describing the evolution of network state x. In particular, the stationary measure of the process $X(t)$ is given by:

$$\pi(x) = \pi(0)\Phi(x)\alpha_1^{x_1} \ldots \alpha_N^{x_N}, \hspace{2cm} [7.14]$$

where $\alpha_i = \lambda_i/\mu_i$ denotes the traffic intensity at queue i and $\Phi(x)$ the inverse of the product of service capacities along any direct path from state x to state 0, as illustrated by Figure 7.5. The mean sojourn times follow from Little's formula.

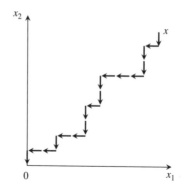

Figure 7.5. *The function Φ(x) is the inverse of the product of service capacities along any path from state x to state 0*

In fact, it can be shown like for Jackson networks that the routing does not change the stationary distribution of the network state as long as the arrival rates at each queue, given by the traffic equations [7.1] and [7.2], are preserved. The partial balance equations are given by:

$$\pi(x)\mu_i m_i(x) = \pi(x - e_i)\nu_i$$

$$+ \sum_{j=1}^{N} \pi(x - e_i + e_j)\mu_j m_j(x - e_i + e_j)p_{ji}, \qquad i = 1, \ldots, N,$$

and:

$$\pi(x)\nu = \sum_{i=1}^{N} \pi(x + e_i)\mu_i m_i(x + e_i)p_i.$$

Replacing π by its expression [7.14], and using the fact that, for any state x:

$$m_i(x) = \frac{\Phi(x - e_i)}{\Phi(x)}, \quad i = 1, \ldots, N : x_i \geq 1, \qquad [7.15]$$

we obtain the traffic equations [7.1] and [7.2], respectively. These results apply to closed networks as well, with a fixed number of customers.

7.7. Kelly networks

We conclude this chapter with Kelly networks, which differ from Jackson networks due to the presence of customer *classes*. We consider N single-server queues under the processor sharing (PS) service discipline. Let K be the number of classes. Class k customers arrive according to a Poisson process of intensity $\nu(k)$ and follow a route of length $L(k)$ in the network, given by the sequence of queues $r(k,1), r(k,2), \ldots, r(k,L(k))$. Service times are exponential with parameter $\mu(k,l)$ at the lth queue visited by class k customers. The load of queue i is thus given by:

$$\bar{\rho}_i = \sum_{k=1}^{K} \nu(k) \sum_{l:r(k,l)=i} \frac{1}{\mu(k,l)}.$$

The key result of Kelly networks is that the stationary distribution π of the network state has the product form. Denoting by $\bar{x}_i$ the total number of customers in queue i and by $\bar{x}$ the vector $(\bar{x}_1, \ldots, \bar{x}_N)$, we have:

$$\pi(\bar{x}) = \prod_{i=1}^{N} (1 - \bar{\rho}_i)\bar{\rho}_i^{\bar{x}_i}, \quad \bar{x} \in \mathbb{N}^N, \qquad [7.16]$$

under the stability condition $\bar{\rho}_i < 1$ for each queue i. Again, the number of customers in each queue is independent of the number of customers in the other queues. Moreover, the distribution of the number of customers in each queue depends on the classes of these customers through their load only.

A simple way to prove equation [7.16] is to observe that the system can be viewed as a Whittle network where each queue corresponds to a certain type of customers, characterized by their class k and the step of their route l. Thus, there are $J = \sum_{k=1}^{K} L(k)$ queues. We denote by x_j the number of customers in queue $j = (k, l)$ and by x the vector $(x_1, \ldots, x_J)$. The total number of customers at queue i in the Kelly network is given by:

$$\bar{x}_i = \sum_{j:(j)=i} x_j.$$

The Whittle network has the following characteristics. Customers arrive at queue $j = (k, 1)$ according to a Poisson process of intensity $\nu(k)$. Service times at queue $j = (k, l)$ are exponential with parameter $\mu(k, l)$. The service discipline at each queue of the Kelly network being PS, the service capacity of queue j of the Whittle network is given by:

$$m_j(x) = \frac{x_j}{\bar{x}_i},$$

with $i = r(j)$. Finally, customers leaving queue $j = (k, l)$ join queue $j' = (k, l+1)$ if $l < L(k)$ and leave the network otherwise, so that the arrival rate of customers at queue j is given by $\nu(k)$. We deduce the traffic intensity at queue $j = (k, l)$:

$$\alpha_j = \frac{\nu(k)}{\mu(k, l)}.$$

We verify that balance condition [7.13] is satisfied; from equation [7.14], the stationary measure of the Markov process $X(t)$ describing the evolution of the number of customers in each queue of the Whittle network is given by:

$$\pi(x) = \pi(0) \prod_{i=1}^{N} \bar{x}_i! \prod_{j=1}^{J} \frac{\alpha_j^{x_j}}{x_j!}, \quad x \in \mathbb{N}^J.$$

Equation [7.16] follows by summation.

7.8. Exercises

1. Feed-forward network

We consider a Jackson network where queues can be indexed in such a way that $p_{ij} = 0$ for all i, j such that $i \geq j$. Find Jackson's result using the departure process from an $M/M/1$ queue (see exercise 1 of Chapter 6).

2. Reversibility

Give a necessary and sufficient condition for the reversibility of the Markov process describing the state of a Jackson network.

3. Observer paradox

Show that, in a Jackson network, the stationary distribution of the state seen by a customer present in a queue is given by the stationary distribution π weighted by the number of customers in this queue. We might consider the PS service discipline and apply the same method as in section 6.10.

4. Migration

A closed Jackson network is split into two parts. Show that the frequency of migration of customers from one part to the other is the same in both directions of migration. Is this true for an open Jackson network?

5. MUSTA property for a closed network

Show that in a closed Jackson network with M customers, the distribution of the network state as seen by a customer after their departure from one queue and just before their arrival in another queue corresponds to the stationary distribution of this network, but with $M - 1$ customers.

6. *Arrival rate in a closed network*

Let Γ_M be the normalization constant associated with the stationary distribution π of the state of a closed Jackson network with M customers, so that:

$$\pi(x) = \Gamma_M \prod_{i=1}^{N} \rho_i^{x_i}, \quad \forall x \in \mathbb{N}^N : \sum_{i=1}^{N} x_i = M.$$

Show that the arrival rate of customers at queue i coming from queue j is given by:

$$\frac{\Gamma_M}{\Gamma_{M-1}} \lambda_j p_{ji}.$$

Deduce the arrival rate of customers at queue i.

7. *Jackson network of infinite-server queues*

We consider a Jackson network in which each queue has an infinite number of servers. Show that this network belongs to the class of Whittle networks. Denoting by $\alpha_i = \lambda_i / \mu_i$ the traffic intensity at queue i, verify that the stationary measure of the Markov process describing the network state is given by:

$$\pi(x) = \pi(0) \frac{\alpha_1^{x_1}}{x_1!} \cdots \frac{\alpha_N^{x_N}}{x_N!}.$$

8. *Multiclass PS queue*

We consider a single-server PS queue receiving a stream of customers belonging to K different classes. Class k customers arrive according to a Poisson process of intensity λ_k and have exponential service times with parameter μ_k. We denote by x_k the number of class k customers, and by $\rho_k = \lambda_k / \mu_k$ their contribution to the load of the queue. Show that this multiclass queue can be viewed as a Whittle network of K

monoclass queues. Deduce the stationary measure of the state of the multiclass queue:

$$\pi(x) = \pi(0) \begin{pmatrix} x_1 + \ldots + x_K \\ x_1, \ldots, x_K \end{pmatrix} \rho_1^{x_1} \ldots \rho_K^{x_K}.$$

What is the stability condition of the queue? Under this condition, verify that the mean number of customers of any class is proportional to the load of this class.

7.9. Solution to the exercises

1. Feed-forward network

Queues can be indexed according to their position in the network, as illustrated by Figure 7.6: queue in position 1 has only external arrivals, queue in position 2 may receive customers coming from outside and from queue in position 1, queue in position 3 may receive customers coming from outside and from queues in positions 1 and 2, etc.

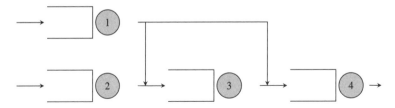

Figure 7.6. *A feed-forward network*

Queue in position 1 is an $M/M/1$ queue; its departure process is thus a Poisson process. Moreover, it follows from the property of subdivision of a Poisson process that the processes of internal arrivals coming from this departure process are mutually independent Poisson processes.

Queue in position 2 thus receives a stream of internal arrivals and a stream of external arrivals, each being an independent Poisson process. By the property of superposition of Poisson processes, the global stream is a Poisson process and queue in position 2 is an $M/M/1$ queue, independent of queue in position 1. Indeed, the departure process of customers from an $M/M/1$ queue until time t is independent of the state of this queue at time t (see exercise 1 of Chapter 6). By induction, all queues behave as mutually independent $M/M/1$ queues, which is precisely Jackson's result.

2. Reversibility

The local balance equations of the Markov process describing the state of a Jackson network are given for all states x by:

$$\forall i, \quad \pi(x + e_i)\mu_i p_i = \pi(x)\nu_i \quad \text{and}$$

$$\forall i, j, \quad \pi(x + e_i)\mu_i p_{ij} = \pi(x + e_j)\mu_j p_{ji}.$$

From equation [7.6], these equations are equivalent to:

$$\forall i, \quad \lambda_i p_i = \nu_i \quad \text{and} \quad \forall i, j, \quad \lambda_i p_{ij} = \lambda_j p_{ji},$$

which are the local balance equations of the Markov chain associated with the random path followed by any customer in the network (see section 7.2). Thus, a Jackson network is reversible if and only if the associated routing process is reversible.

3. Observer paradox

We are interested in the state of queue 1 as seen by an observer. The activity periods of the observer have exponential durations with parameter γ. We build a Markov process $(X(t), Y(t))$, as shown in section 6.10, $X(t)$ representing the number of standard customers in each queue and $Y(t)$ the activity state of the observer at time t. The stationary measure of this process is given by:

$$\pi(x,0) = \pi(0,0) \prod_{i=1}^{N} \rho_i^{x_i} \quad \text{and}$$

$$\pi(x,1) = \pi(0,0)(x_1+1)\beta_1 \prod_{i=1}^{N} \rho_i^{x_i}, \ x \in \mathbb{N}^N,$$

with $\beta_1 = \gamma/\mu_1$. We indeed verify that $\pi(.,0)$ satisfies the partial balance equations [7.7] and [7.8] that $\pi(.,1)$ satisfies the following modified partial balance equations:

$$\pi(x,1)\mu_1 \frac{x_1}{x_1+1} = \pi(x-e_1,1)\nu_1 + \sum_{j \neq 1} \pi(x-e_1+e_j,1)\mu_j p_{j1}$$

$$+ \pi(x,1)\mu_1 \frac{x_1}{x_1+1} p_{11},$$

$$\pi(x,1)\mu_i \mathbf{1}(x_i \geq 1) = \pi(x-e_i)\nu_i + \sum_{j \neq 1} \pi(x-e_i+e_j,1)\mu_j p_{ji}$$

$$+ \pi(x-e_i+e_1,1)\mu_1 \frac{x_1}{x_1+1} p_{1i}, \quad i \neq 1,$$

and that, regarding the observer:

$$\gamma\pi(x,0) = \frac{\mu_1}{x_1+1}\pi(x,1).$$

The balance equations follow by summation. The probability that the observer sees the system in state x (including the observer) when the observer is active is $\pi'(x) \propto \pi(x-e_1,1)$, that is:

$$\pi'(x) = \pi'(0)x_1 \prod_{i=1}^{N} \rho_i^{x_i}, \ x \in \mathbb{N}^N.$$

This is the stationary distribution of the initial network, without observer, weighted by the number of customers in queue 1.

4. Migration

We split the set of queues into two sets $\mathcal{A}$ and $\mathcal{B}$ and write equation [7.12] so that, for any queue $i \in \mathcal{A}$ and all states x such that $x_i > 0$:

$$\sum_{j \in \mathcal{A}} \pi(x) \mu_i p_{ij} + \sum_{j \in \mathcal{B}} \pi(x) \mu_i p_{ij}$$

$$= \sum_{j \in \mathcal{A}} \pi(x - e_i + e_j) \mu_j p_{ji} + \sum_{j \in \mathcal{B}} \pi(x - e_i + e_j) \mu_j p_{ji}.$$

Summing this equation over all states x and all queues in $\mathcal{A}$, we obtain:

$$\sum_{x \in \mathcal{X}} \sum_{i \in \mathcal{A}} 1(x_i > 0) \sum_{j \in \mathcal{A}} \pi(x) \mu_i p_{ij} + \sum_{x \in \mathcal{X}} \sum_{i \in \mathcal{A}} 1(x_i > 0) \sum_{j \in \mathcal{B}} \pi(x) \mu_i p_{ij}$$

$$= \sum_{x \in \mathcal{X}} \sum_{j \in \mathcal{A}} 1(x_j > 0) \sum_{i \in \mathcal{A}} \pi(x) \mu_j p_{ji} + \sum_{x \in \mathcal{X}} \sum_{j \in \mathcal{B}} 1(x_j > 0) \sum_{i \in \mathcal{A}} \pi(x) \mu_j p_{ji}.$$

After simplification, we obtain the sought result:

$$\sum_{x \in \mathcal{X}} \sum_{i \in \mathcal{A}} 1(x_i > 0) \sum_{j \in \mathcal{B}} \pi(x) \mu_i p_{ij} = \sum_{x \in \mathcal{X}} \sum_{j \in \mathcal{B}} 1(x_j > 0) \sum_{i \in \mathcal{A}} \pi(x) \mu_j p_{ji}.$$

This result is not true, in general, for an open network. It suffices to consider two queues, 1 and 2, such that customers arrive at rate λ at queue 1 and then move to queue 2 before leaving the network as represented in Figure 7.7. If both the queues have a service rate $\mu > \lambda$ then the customers move from queue 1 to queue 2 at rate λ but no customer moves from queue 2 to queue 1. It can, in fact, be shown that the result holds for open networks whose routing process is reversible; see exercise 2.

Figure 7.7. *Two queues in tandem*

5. MUSTA property for a closed network

Let π be the stationary measure of the Markov process describing the state of the network:

$$\pi(x) = \prod_{i=1}^{N} \rho_i^{x_i}, \quad \forall x \in \mathbb{N}^N : \sum_{i=1}^{N} x_i = M.$$

For any state x such that $|x| = M-1$, the frequency of the event "a customer moving from queue i to queue j sees the network in state x just before her arrival at queue j" is proportional to $\pi(x + e_i)\mu_i p_{ij}$. From the formula of conditional transitions of section 5.11, the probability that a customer moving from queue i to queue j sees the network in state x before their arrival is given by:

$$\pi'(x) = \frac{\pi(x + e_i)\mu_i p_{ij}}{\sum_{y:|y|=M-1} \pi(y + e_i)\mu_i p_{ij}}.$$

From the expression of the stationary measure, we have $\pi(x + e_i)\mu_i = \pi(x)\lambda_i$ so that:

$$\pi'(x) = \frac{\pi(x)}{\sum_{y:|y|=M-1} \pi(y)}.$$

From equation [7.11], this is the stationary distribution of the same network, but with $M - 1$ customers.

6. Arrival rate in a closed network

From section 5.8, the frequency of transitions corresponding to the movement of a customer from queue j to queue i is given by:

$$\sum_{x:|x|=M-1} \pi(x + e_j)\mu_j p_{ji}.$$

Using the fact that $\pi(x + e_j)\mu_j = \pi(x)\lambda_j$, we obtain:

$$\sum_{x:|x|=M-1} \pi(x)\lambda_j p_{ij} = \frac{\Gamma_M}{\Gamma_{M-1}}\lambda_j p_{ij}.$$

The arrival rate at queue i follows by summation on j. Using traffic equations [7.10], we deduce:

$$\frac{\Gamma_M}{\Gamma_{M-1}}\lambda_i.$$

7. Jackson network of infinite-server queues

The considered network is a Whittle network because the service capacities, given by $m_i(x) = x_i$ for queue i, satisfy balance equation [7.13]. The stationary distribution follows from the fact that:

$$\Phi(x) = \frac{1}{x_1!\ldots x_N!}.$$

8. Multiclass PS queue

For each class k, we define a queue of service rate $\mu_k m_k(x)$, with:

$$m_k(x) = \frac{x_k}{x_1 + \ldots + x_K}.$$

Balance equation [7.13] holds. Moreover, we have:

$$\Phi(x) = \binom{x_1 + \ldots + x_K}{x_1, \ldots, x_K}.$$

The stationary measure follows from equation [7.14]:

$$\pi(x) = \pi(0)\binom{x_1 + \ldots + x_K}{x_1, \ldots, x_K}\rho_1^{x_1}\ldots\rho_K^{x_K}.$$

The queue is stable if and only if $\rho < 1$, with $\rho = \rho_1 + \ldots + \rho_K$. The mean number of class k customers is then given by:

$$\mathrm{E}(X_k) = \sum_{x : x_k \geq 1} x_k \pi(x),$$

$$= \pi(0) \sum_{x : x_k \geq 1} (x_1 + \ldots + x_K) \binom{x_1 + \ldots + x_K - 1}{x_1, \ldots, x_k - 1, \ldots, x_K} \rho_1^{x_1} \ldots \rho_K^{x_K},$$

$$= \rho_k \sum_x (x_1 + \ldots + x_K + 1) \pi(x),$$

which is indeed proportional to the load ρ_k class k.

Chapter 8

Circuit Traffic

The Erlang model was developed a century ago to dimension the first telephone networks. Today, it is still a reference in the field of telecommunications. The model and its extensions are instrumental in the performance analysis of the so-called "loss" systems whose resources are reserved and incoming clients are rejected in case of congestion.

8.1. Erlang's model

Consider a link consisting of m circuits. Each telephone call requires one circuit. Calls arrive according to a Poisson process of intensity λ and have exponential durations with parameter μ (this assumption is, in fact, not essential due to the insensitivity property, see exercise 4 in section 8.9). When all circuits are occupied, incoming calls are blocked and lost. We are interested in the *blocking rate*, that is, the probability that an incoming call is blocked. A crucial parameter is the traffic intensity, defined as the product of the call arrival rate and the mean call duration:

$$\alpha = \frac{\lambda}{\mu}. \qquad [8.1]$$

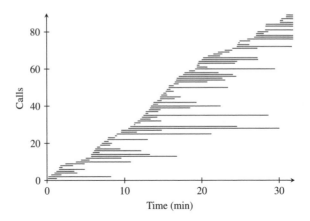

Figure 8.1. *Randomness of telephone traffic*

This dimensionless quantity, generally expressed in *erlangs* (symbol E), corresponds to the mean number of calls in the absence of blocking, that is when $m = \infty$. The system is then equivalent to an $M/M/\infty$ queue, whose steady state has a Poisson distribution with mean α (see section 6.4). The randomness of traffic is illustrated in Figure 8.1 for an arrival rate of 2.5 calls per minute and a mean call duration of 4 minutes, corresponding to a traffic intensity of $\alpha = 10\,\mathrm{E}$.

The link load is the ratio of traffic intensity to link capacity:

$$\rho = \frac{\alpha}{m} = \frac{\lambda}{m\mu}.$$

8.2. Erlang's formula

The Erlang model is nothing more than an $M/M/m/m$ queue. In particular, the stationary distribution of the number of ongoing calls is given by:

$$\forall x = 0, 1, \ldots, m, \quad \pi(x) = \frac{\frac{\alpha^x}{x!}}{1 + \alpha + \frac{\alpha^2}{2} + \ldots + \frac{\alpha^m}{m!}}. \qquad [8.2]$$

This is a truncated Poisson distribution, as illustrated in Figure 8.2. According to the Poisson Arrivals See Time Averages (PASTA) property, each call sees the system in steady state at its arrival and thus is blocked with probability $\pi(m)$. We deduce the blocking rate:

$$B = \frac{\frac{\alpha^m}{m!}}{1 + \alpha + \frac{\alpha^2}{2} + \ldots + \frac{\alpha^m}{m!}}. \qquad [8.3]$$

This is the Erlang formula, published in 1917.

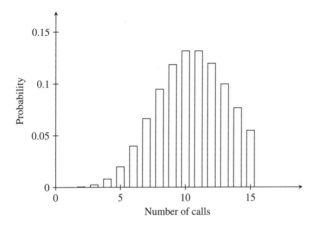

Figure 8.2. *Distribution of the number of calls for $\alpha = 10\,E$ and $m = 15$ circuits*

Figure 8.3 illustrates the blocking rate with respect to the link load $\rho = \alpha/m$ for different values of m. Note that, at constant load, the blocking rate decreases with capacity. These economies of scale can be explained by the lower traffic fluctuations (see exercise 1 in section 8.9) and can be proved using the integral form of the Erlang formula (see exercise 2). In the limit, we get the loss rate of a "fluid" model, without any traffic fluctuations: null if $\rho < 1$ and equal to the fraction of traffic in excess $(\rho - 1)/\rho$ otherwise; this corresponds to the case $m = \infty$ in Figure 8.3.

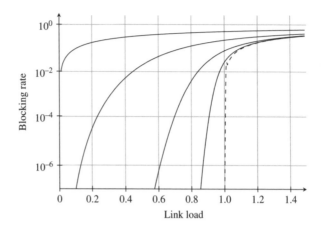

Figure 8.3. *The Erlang formula (m = 1, 10, 100, 1000, ∞,*
from top to bottom)

The success of the Erlang formula is mainly due to its simplicity and robustness: it depends on the traffic characteristics through the traffic intensity α only. In particular, it holds for any distribution of call durations (see exercise 4). Erlang's model is said to be "insensitive". The only critical assumption is the Poisson process of call arrivals, which is satisfied when calls are generated by a large number of users (see section 3.7); for a small number of users, the Engset model described below applies.

The numerical computation of Erlang's formula can be troublesome[1] for large values of m. In this case, it is preferable to calculate the inverse of the blocking rate $I(m)$ for m circuits. Using equation [8.3], we get the following recursive formula:

$$I(m) = 1 + \frac{m}{\alpha} I(m - 1), \quad \text{with} \quad I(0) = 1. \qquad [8.4]$$

The limiting value I of the inverse of the blocking rate when the capacity m tends to infinity and the link load $\rho = \alpha/m$

---

1 For example, the number 1,000! is of the same order as 10^{250}.

is kept constant can be obtained by solving the corresponding equation: $I = 1 + I/\rho$. We get $I = \infty$ for $\rho < 1$, which corresponds to a null blocking rate, and $I = \rho/(\rho - 1)$ otherwise, which corresponds to the blocking rate $(\rho - 1)/\rho$.

8.3. Engset's formula

The Engset model is very similar to the Erlang model; the only difference is that calls are not generated according to a Poisson process but by the activity of a fixed number of users, denoted by K. Specifically, each user generates a continuous sequence of calls separated by random idle periods, as illustrated in Figure 8.4.

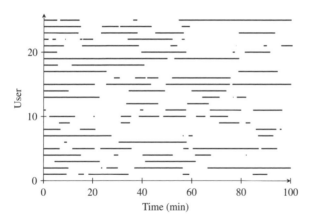

Figure 8.4. *Traffic generated by $K = 25$ users with per-user activity rate $a = 0.4$*

Call durations are exponential with parameter μ and idle periods are exponential with parameter ν. Denote by β the ratio ν/μ. In the absence of blocking, the traffic intensity generated by each user is the fraction of time where the user is active. The evolution of the activity state of each user is then described by a Markov process whose transition graph is shown in Figure 8.5, where 0 refers to the idle state and 1 to the active state. In particular, the probability that the user is active is given by:

$$a = \frac{\nu}{\nu + \mu} = \frac{\beta}{1 + \beta}.$$

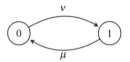

Figure 8.5. *Transition graph of the Markov process describing the state of a user*

8.3.1. *Model without blocking*

We first consider the evolution of the number x of ongoing calls in the absence of blocking. This is a birth–death process whose transition graph is shown in Figure 8.6.

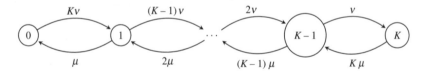

Figure 8.6. *Transition graph associated with the Engset model without blocking*

We get the stationary measure:

$$\pi(x) \propto \binom{K}{x}\beta^x, \quad x = 0, 1, \ldots, K,$$

and by normalization,

$$\pi(x) = \binom{K}{x}a^x(1 - a)^{K-x}.$$

Thus, in the absence of blocking, the number of ongoing calls has a *binomial* distribution in steady state. An example of such a distribution is illustrated in Figure 8.7.

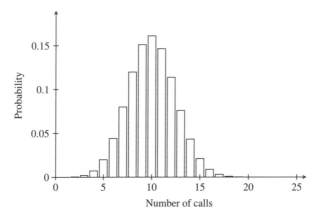

Figure 8.7. *Distribution of the number of calls for $K = 25$ users with per-user activity rate $a = 0.4$*

8.3.2. *Model with blocking*

In the presence of admission control, any user whose call is blocked stays idle for a new period of exponential duration with parameter ν. For m circuits, with $m < K$, we obtain the stationary distribution from truncation (see section 5.15):

$$\pi(x) \propto \binom{K}{x}\beta^x, \quad x = 0, 1, \ldots, m.$$

The blocking rate is obtained by calculating the ratio of the frequency of blocked calls to the frequency of incoming calls:

$$B = \frac{(K - m)\nu\pi(m)}{K\nu\pi(0) + (K - 1)\nu\pi(1) + \ldots + (K - m)\nu\pi(m)}.$$

We get the Engset formula:

$$B = \frac{\binom{K-1}{m}\beta^m}{1 + \binom{K-1}{1}\beta + \binom{K-1}{2}\beta^2 + \cdots + \binom{K-1}{m}\beta^m}. \qquad [8.5]$$

Figure 8.8 illustrates the blocking rate with respect to the link load $\rho = \alpha/m$, with $\alpha = K\beta$. When the number of users K increases, at constant load, the blocking rate tends to the Erlang formula. Indeed, the superposition of a large number of independent stationary point processes of low intensity tends to a Poisson process (see section 3.7). Moreover, at any given load, the Engset formula always results in a lower blocking rate than the Erlang formula: the Erlang formula is the worst case and can always be used for dimensioning purposes. This explains why the Erlang formula, simpler and more conservative than the Engset formula, is commonly used in practice.

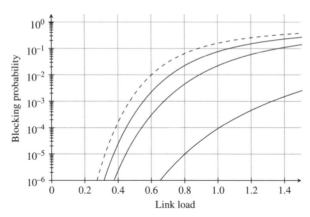

Figure 8.8. *The Engset formula (m = 20, K = 25, 50, 100, ∞, from bottom to top)*

Finally, we have a recursive formula similar to equation [8.4] for the numerical computation of the Engset formula:

$$I(m) = 1 + \frac{m}{(K-m)\beta}I(m-1), \quad \text{with} \quad I(0) = 1.$$

8.4. Erlang's waiting formula

We have so far considered pure loss systems only. In order to decrease the blocking rate, it might be interesting to queue the incoming calls when all circuits are occupied, hoping that a circuit frees quickly. This is a common practice in call centers, for instance. Consider again the Erlang model but assume that the system is now equipped with an infinite queue in order to collect the blocked calls. These are served in FIFO order. This system is referred to as *Erlang's waiting model*. It corresponds to the $M/M/m$ queue, whose stationary measure is given by:

$$\pi(x) = \pi(0)\frac{\alpha^x}{x!}, \qquad \forall x = 0, 1, \ldots, m$$

[8.6]

$$\pi(x) = \pi(0)\frac{\alpha^m}{m!}\rho^{x-m}, \quad \forall x > m.$$

The system is stable if and only if $\rho < 1$, which is assumed in the following subsections.

8.4.1. *Waiting probability*

In view of the PASTA property, an incoming call must wait with probability:

$$Q = \sum_{x \geq m} \pi(x) = \frac{\frac{\alpha^m}{m!}\frac{1}{1-\rho}}{1 + \alpha + \frac{\alpha^2}{2} + \ldots + \frac{\alpha^m}{m!} + \frac{\alpha^m}{m!}\frac{\rho}{1-\rho}}.$$

This is Erlang's waiting formula, shown in Figure 8.9 with respect to the link load ρ for different values of capacity m. Again, we observe that performance improves with capacity, at any given load. This is due to the lower traffic fluctuations.

It is worth noting that the Erlang waiting formula is linked to the corresponding Erlang loss formula B through the following relationship:

$$Q = \frac{B}{1 - \rho + \rho B}.$$

[8.7]

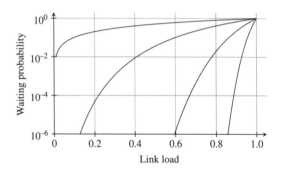

Figure 8.9. *The Erlang waiting formula (m = 1, 10, 100, 1000, from top to bottom)*

In particular, the recursive formula [8.4] enables us to calculate efficiently this quantity.

8.4.2. *Mean waiting time*

Another interesting performance metric is the mean waiting time of queued calls. In view of equation [8.6], the number of ongoing calls X satisfies:

$$P(X = x \mid X \geq m) = \frac{P(X = x)}{P(X \geq m)} = (1 - \rho)\rho^{x-m}, \quad \forall x \geq m.$$

Thus, given that an incoming call is queued, the number of queued calls $X - m$ has the same distribution as the number of customers in an $M/M/1$ queue of load ρ. In particular, the mean conditional number of queued calls is given by:

$$E(X - m \mid X \geq m) = \frac{\rho}{1 - \rho}.$$

Since calls leave the system at rate $m\mu$, we deduce the mean conditional waiting time:

$$\delta = (E(X - m \mid X \geq m) + 1)\frac{1}{m\mu} = \frac{1}{m\mu - \lambda}. \qquad [8.8]$$

Contrary to the Erlang loss model, this performance result is sensitive to the distribution of call durations. Like the $M/G/1$ queue (see Pollaczek–Kinchin's formula in section 6.9), the mean waiting time increases with the *variance* of call durations, at constant mean.

8.5. The multiclass Erlang model

We now introduce different extensions of the Erlang model, accounting for both heterogeneous requests and multiple links. These extensions were developed when the Integrated Services Digital Network (ISDN) emerged, when the telephone network infrastructure was used to transfer data flows over several circuits in parallel. They are instrumental in the performance evaluation of any communication network or computer system involving resource reservation.

Consider N call classes. Each class i call requires c_i circuits, as illustrated in Figure 8.10. The system consists of m circuits. Any call that cannot be served is blocked and lost. We are interested in the blocking rate of each class of calls.

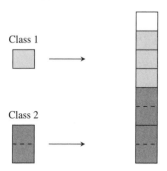

Figure 8.10. *The multiclass Erlang model with $N = 2$ classes*
($m = 8$, $c = (1, 2)$)

Assume that class i calls arrive according to a Poisson process of intensity λ_i and have exponential durations with parameter μ_i (as for the Erlang model, results are, in fact, insensitive, that is independent of the distribution of call durations beyond the mean). Denote by $\alpha_i = \lambda_i/\mu_i$ the corresponding traffic intensity in erlangs and by α the total traffic intensity in circuits. Since a class i call requires c_i circuits, we get:

$$\alpha = \sum_{i=1}^{N} \alpha_i c_i.$$

The link load is the ratio of traffic intensity to capacity:

$$\rho = \frac{\alpha}{m}.$$

Let x_i be the number of class i calls. Denote by x and c the respective vectors $(x_1,\ldots,x_N)$ and $(c_1,\ldots,c_N)$. The set of admissible states is defined by the capacity constraint:

$$x.c \equiv x_1 c_1 + \ldots + x_N c_N \le m.$$

In the absence of blocking (when $m = \infty$), the system reduces to N independent birth–death processes. This is a reversible process (see section 5.16) with stationary measure:

$$\pi(x) = \pi(0)\frac{\alpha_1^{x_1}}{x_1!} \cdots \frac{\alpha_N^{x_N}}{x_N!}.$$

The stationary distribution of the original process (with $m < \infty$) is obtained by truncation and normalization:

$$\pi(x) = \pi(0)\frac{\alpha_1^{x_1}}{x_1!} \cdots \frac{\alpha_N^{x_N}}{x_N!}, \quad x.c \le m.$$

We get the blocking rate of class i calls, thanks to the PASTA property. This is the probability that the system is in a blocking state that is unable to accept any incoming class i call:

$$B_i = \sum_{x:m-c_i < x.c \le m} \pi(x). \tag{8.9}$$

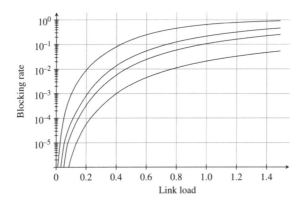

Figure 8.11. *Blocking rates of $N = 4$ classes of calls in the multiclass Erlang model ($m = 100$, $c = (1, 5, 10, 30)$)*

Note that the higher the number c_i of requested circuits, the higher the blocking rate. The results are illustrated in Figure 8.11 for $N = 4$ call classes with a homogeneous traffic distribution, that is, with $\alpha_1 c_1 = \alpha_2 c_2 = \alpha_3 c_3 = \alpha_4 c_4$.

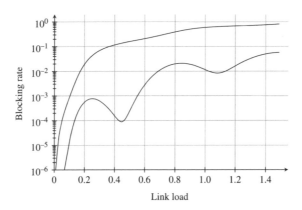

Figure 8.12. *Example of a non-monotonic blocking rate ($m = 100$, $c = (1, 30)$)*

Figure 8.12 illustrates the fact that, surprisingly, the blocking rate is generally not a monotonic function of the link load. This is due to the fact that increasing the load might penalize some call classes (those requesting a high number of circuits) and, in turn, benefit to some other call classes (those requesting a low number of circuits).

8.6. Kaufman–Roberts formula

The number of states to consider in the calculation of the blocking probabilities [8.9] is *exponential* in the number of call classes N. The following recursive formula makes the calculation *linear* in N. The idea is to calculate the distribution of the link occupancy $n = x.c$, as illustrated in Figure 8.13. Let:

$$f(n) = \sum_{x:x.c=n} \frac{\alpha_1^{x_1}}{x_1!} \cdots \frac{\alpha_N^{x_N}}{x_N!},$$

we have:

$$f(n) = \frac{1}{n} \sum_{i=1}^{N} \alpha_i c_i f(n - c_i), \qquad [8.10]$$

with $f(0) = 1$ and the convention that $f(n) = 0$ if $n < 0$. This is the *Kaufman–Roberts formula*. The blocking rate of class i calls then follows as:

$$B_i = \frac{\sum_{m-c_i < n \le m} f(n)}{\sum_{0 \le n \le m} f(n)}.$$

To prove the recursive formula [8.10], we write:

$$f(n) = \sum_{x:x.c=n} \frac{\alpha_1^{x_1}}{x_1!} \cdots \frac{\alpha_N^{x_N}}{x_N!}$$

$$= \sum_{x:x.c=n} \frac{x.c}{n} \times \frac{\alpha_1^{x_1}}{x_1!} \cdots \frac{\alpha_N^{x_N}}{x_N!}$$

$$= \sum_{x:x.c=n} \frac{1}{n} \sum_{i:x_i \ge 1} \alpha_i c_i \frac{\alpha_1^{x_1}}{x_1!} \cdots \frac{\alpha_i^{x_i-1}}{(x_i - 1)!} \cdots \frac{\alpha_N^{x_N}}{x_N!}$$

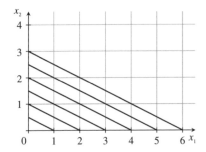

Figure 8.13. *Summation over states x such that x.c is constant*
(C = 6, c = (1, 2))

Finally, we get:

$$f(n) = \frac{1}{n} \sum_{i=1}^{N} \alpha_i c_i \sum_{x:x.c=n-c_i} \frac{\alpha_1^{x_1}}{x_1!} \cdots \frac{\alpha_N^{x_N}}{x_N!},$$

$$= \frac{1}{n} \sum_{i=1}^{N} \alpha_i c_i f(n - c_i).$$

8.7. Network models

We now consider the impact of multiple links. For the sake of simplicity, we assume that each call requires a single circuit; the results extend, as in the previous section, to the case where calls require an arbitrary number of circuits. With this assumption, calls differ only through their route in the network. A call is accepted if it gets a circuit on each link on its route. Otherwise, it is blocked and lost. We are interested in the blocking rate of each class of calls.

Let L be the number of links and N the number of call classes. Link l contains m_l circuits. The route of a class i call is a subset of the links $R_i \subset \{1, \ldots, L\}$. In the network of Figure 8.14, for instance, there are $N = 5$ call classes with routes $R_1 = \{1,2\}$, $R_2 = \{1,4\}$, $R_3 = \{3\}$, $R_4 = \{3,4\}$, and $R_5 = \{2\}$. As usual, we assume that calls arrive according to Poisson processes and have exponential durations. We denote by α_i the traffic intensity of class i calls in erlangs.

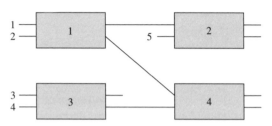

Figure 8.14. *A network with $L = 4$ links and $N = 5$ call classes*

Let x_i be the number of class i calls. Denote by x the vector $(x_1, \ldots, x_N)$. The set of admissible states $\mathcal{X}$ is given by the capacity constraints:

$$\sum_{i:l \in R_i} x_i \leq m_l, \quad l = 1, \ldots, L.$$

In the absence of blocking (i.e. for $m_1 = \ldots = m_L = \infty$), the system corresponds to N independent birth–death processes. This is a reversible Markov process (see section 5.16) with stationary measure:

$$\pi(x) = \pi(0)\frac{\alpha_1^{x_1}}{x_1!} \ldots \frac{\alpha_N^{x_N}}{x_N!}.$$

The stationary distribution of the original process (with $m_1 < \infty, \ldots, m_L < \infty$) is obtained by truncation and normalization:

$$\pi(x) = \pi(0)\frac{\alpha_1^{x_1}}{x_1!} \ldots \frac{\alpha_N^{x_N}}{x_N!}, \quad x \in \mathcal{X}.$$

Thanks to the PASTA property, the blocking rate of class i calls is equal to the probability that the system is in a blocking state, that is, unable to accept any incoming class i call:

$$B_i = \sum_{x \in \mathcal{X}: x + e_i \notin \mathcal{X}} \pi(x).$$

8.8. Decoupling approximation

Like the multiclass Erlang model, the computation complexity of the blocking rates might be excessive when N is large. A usual approximation consists of considering the impact of each link individually.

Specifically, denote by b_l the blocking rate associated with isolated link l. This is given by the Erlang formula with capacity m_l and traffic intensity $\sum_{i:l \in R_i} \alpha_i$. The class i blocking rate is then calculated, thanks to the expression:

$$B_i \approx 1 - \prod_{l \in R_i} (1 - b_l).$$

The approximation consists of assuming that the blocking events on the different links of the route are independent. It is conservative (the actual blocking rate is lower), as illustrated in Figure 8.16 for class 1 calls in the network of Figure 8.15, with the same traffic intensity for both the classes. Note that the approximation is very good for low blocking rates.

Figure 8.15. *A network with $L = 2$ links and $N = 2$ call classes*

8.9. Exercises

1. Traffic fluctuations
Prove that, in the absence of blocking, the relative fluctuations of the number of ongoing calls decrease with the

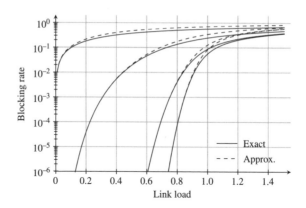

Figure 8.16. *Blocking rate of class 1 calls in the network of Figure 8.15*
($m_1 = 1, 10, 100, 250$, from top to bottom, $m_2 = 2m_1$)

traffic intensity. Calculate the coefficient of variation (ratio of
the standard deviation to the mean) of the number of calls.

2. *Integral form of the Erlang formula*
    Using recursive formula [8.4], prove that the Erlang
    formula satisfies:

$$\frac{1}{B} = \int_0^\infty \left(\frac{t}{\alpha} + 1\right)^m e^{-t}\, \mathrm{d}t.$$

Deduce that the blocking rate decreases with the link capacity
m at constant load $\rho = \alpha/m$.

3. *Admission rate*
    Denote by λ' the arrival rate of admitted calls in the
Erlang model. Prove that $\lambda' = \lambda(1 - B)$, where B is the
corresponding blocking rate, and that $\lambda'/\mu \leq m$ using Little's
formula. Deduce that, when $\rho > 1$, the blocking rate is always
higher than the loss rate of the corresponding fluid model,
namely $(\rho - 1)/\rho$.

4. *Insensitivity of the Erlang formula*
    Consider the Erlang model with hyperexponential call
durations: the duration of each call has an exponential

distribution with parameter μ_1 with probability p_1 and an exponential distribution with parameter μ_2 with probability p_2, with $p_1 + p_2 = 1$. The mean call duration is then equal to $\sigma = p_1/\mu_1 + p_2/\mu_2$ and the traffic intensity is $\alpha = \lambda\sigma$. Prove that the blocking rate is given by Erlang's formula.

5. *Engset's formula*
Express the blocking rate of the Engset model with respect to the mean number of ongoing calls (we might use Little's formula). Deduce a lower bound. Prove the result of exercise 3 by considering the limit when $K \to \infty$, $\beta \to 0$, the link load tending to some positive constant ρ.

6. *Mean conditional waiting time*
Find the mean conditional waiting time equation [8.8] in the Erlang waiting model using Little's formula.

7. *Link occupancy*
Calculate the mean number of occupied circuits in the Erlang waiting model.

8. *Waiting and loss model*
Consider a mix of the Erlang loss model and the Erlang waiting model: when all circuits are occupied, incoming calls are queued but the queue is limited to n calls. Calculate the blocking probability, the waiting probability, and the mean waiting time of queued calls.

9. *Call center with waiting and impatience*
Consider a call center where calls arrive according to a Poisson process and have exponential durations. When all agents are busy, incoming calls are queued. Assume that, for each queued call, the user patience duration is exponential. At the end of this patience duration, the call is abandoned and lost. Calculate the waiting probability and the rate of abandonments.

10. Video on demand

A video-on-demand server provides low-definition and high-definition video streams. These correspond to respective throughputs of 2 and 10 Mbit/s. Assume that 60% of the traffic is in high definition. Given that the server can provide a total throughput of 100 Mbit/s and that an admission control prevents exceeding this limit, calculate the blocking rate of each type of video for a server load of 50%.

11. The generalized Engset model

Consider a link of m circuits shared by K users. User k has exponential call durations with parameter μ_k and exponential idle periods with parameter ν_k; in case of call blocking, a user enters a new idle period as if the calls were completed immediately. Let $x_k = 1$ if user k has a call in progress and $x_k = 0$ otherwise. Show that the corresponding vector x has the stationary distribution:

$$\pi(x) = \pi(0) \prod_{k=1}^{K} \beta_k^{x_k}, \quad x_1 + \ldots + x_K \leq m,$$

where $\beta_k = \nu_k/\mu_k$. Deduce that the blocking probability of each user is independent of his/her traffic intensity. Which user has the highest blocking probability?

8.10. Solutions to the exercises

1. Traffic fluctuations

We consider the Erlang model without blocking. This is an $M/M/\infty$ queue. The number of ongoing calls has a Poisson distribution with mean α. The coefficient of variation is given by $1/\sqrt{\alpha}$, which decreases with α.

2. Integral form of the Erlang formula

Let:

$$J(m) = \int_0^\infty \left(\frac{t}{\alpha} + 1\right)^m e^{-t}\, \mathrm{d}t.$$

Noting that $J(0) = 1$, and by an integration by parts:

$$\forall m \geq 1, \quad J(m) = 1 + \frac{m}{\alpha} J(m-1),$$

we deduce from equation [8.4] that $I(m) = J(m)$ for all $m \geq 0$. Since the function:

$$m \mapsto \left(\frac{t}{\rho m} + 1 \right)^m,$$

increases with m for all $t \geq 0$, the blocking rate decreases with m at fixed load ρ.

3. *Admission rate*

The equality $\lambda' = \lambda(1-B)$ follows from exercise 5 in section 6.11. For each admitted call, the sojourn time in the queue is equal to the service time. By Little's formula, we get $\lambda'/\mu = E(X) \leq m$. Since $\rho = \lambda/(m\mu)$, we deduce that $1 - B \leq 1/\rho$ and the result follows.

4. *Insensitivity of the Erlang formula*

Without the capacity constraint, the model corresponds to two parallel $M/M/\infty$ queues with respective arrival rates λp_1 and λp_2 and respective service rates μ_1 and μ_2 (we use here the property of subdivision of a Poisson process, see section 3.6). The states of these queues consist of two independent reversible Markov processes, with respective stationary measures:

$$\pi_1(x_1) = \pi_1(0)\frac{\alpha_1^{x_1}}{x_1!} \quad \text{and} \quad \pi_2(x_2) = \pi_2(0)\frac{\alpha_2^{x_2}}{x_2!}.$$

The joint state of these two queues defines a reversible Markov process whose stationary measure is the product measure $\pi(x) = \pi_1(x_1)\pi_2(x_2)$ (see section 5.16). The stationary measure π' of the original process, with capacity constraint, is obtained by truncation (see section 5.15):

$$\pi'(x) = \pi'(0)\frac{\alpha_1^{x_1}}{x_1!}\frac{\alpha_2^{x_2}}{x_2!}, \quad x_1 + x_2 \leq m.$$

Normalization gives:

$$\pi'(0) \sum_{x_1, x_2 : x_1 + x_2 \le m} \frac{\alpha_1{}^{x_1} \alpha_2{}^{x_2}}{x_1! \, x_2!} = 1.$$

Finally, we get:

$$\pi'(0) = \left(\sum_{n=0}^{m} \frac{1}{n!} \sum_{x_1, x_2 : x_1 + x_2 = n} \binom{n}{x_1} \frac{\alpha_1{}^{x_1} \alpha_2{}^{x_2}}{x_1! \, x_2!} \right)^{-1},$$

$$= \frac{1}{1 + \alpha + \frac{\alpha^2}{2} + \ldots + \frac{\alpha^m}{m!}}.$$

The distribution of the total number of ongoing calls is given by equation [8.2] and thus coincides with that of the Erlang model with exponential call durations. According to the PASTA property, the blocking rate is the probability that the m circuits are occupied and is given by Erlang's formula.

5. Engset's formula

Let λ be the call arrival rate and λ' be the arrival rate of admitted calls. By applying Little's formula to the ongoing calls and the idle users, respectively, we get:

$$\frac{\lambda'}{\mu} = \mathrm{E}(X) \quad \text{and} \quad \frac{\lambda}{\nu} = K - \mathrm{E}(X).$$

We deduce the blocking rate:

$$B = 1 - \frac{\lambda'}{\lambda} = 1 - \frac{\mathrm{E}(X)}{\beta(K - \mathrm{E}(X))}.$$

Since $\mathrm{E}(X) \le m$, we get the lower bound:

$$B \ge 1 - \frac{m}{\beta(K - m)}.$$

In the limit $K \to \infty$ and $\beta \to 0$ with $K\beta = \alpha$, this lower bound becomes $1 - 1/\rho$, which corresponds to the bound obtained in exercise 3 for the Erlang model.

6. Mean conditional waiting time

Since only a fraction Q of calls is queued, Little's formula applied to these calls gives:

$$\lambda Q \delta = E(\max(X - m, 0)).$$

According to equation [8.6]:

$$E(\max(X - m, 0)) = \sum_{x > m} \pi(0) \frac{\alpha^m}{m!} \rho^{x-m} (x - m) = Q \frac{\rho}{1 - \rho}.$$

We deduce the mean conditional waiting time:

$$\delta = \frac{1}{m\mu - \lambda}.$$

7. Link occupancy

Since the mean call duration is $1/\mu$, it follows from Little's formula applied to the m servers that the mean number of occupied servers is equal to λ/μ, that is, $m\rho$ (see exercise 6 in section 6.11).

8. Waiting and loss model

This model corresponds to an $M/M/m/m + n$ queue. The associated Markov process is the restriction of the Markov process of the Erlang waiting model to the state space $\{0, 1, \ldots, m + n\}$. Since this process is reversible, we get the stationary distribution by truncation and normalization:

$$\pi(x) = \pi(0) \frac{\alpha^x}{x!}, \qquad \forall x = 0, 1, \ldots, m,$$

$$\pi(x) = \pi(0) \frac{\alpha^m}{m!} \rho^{x-m}, \qquad \forall x = m + 1, \ldots, m + n,$$

with:

$$\pi(0) = \frac{1}{1 + \alpha + \ldots + \frac{\alpha^m}{m!} + \frac{\alpha^m}{m!} (\rho + \ldots + \rho^{m+n})}.$$

Thanks to the PASTA property, the blocking rate is $B = \pi(m + n)$ and the waiting probability is $Q = \pi(m) + \ldots + \pi(m+n-1)$. Given that an incoming call is queued, the mean number of queued calls is given by:

$$\mathrm{E}(X - m \mid m \leq X < m + n) = \frac{\rho + 2\rho + \ldots + (n-1)\rho^{n-1}}{1 + \rho + \ldots + \rho^{n-1}}.$$

Since calls leave the system at rate $m\mu$, we deduce the mean conditional waiting time:

$$\delta = \frac{1}{m\mu}(\mathrm{E}(X - m \mid m \leq X < m + n) + 1)$$

$$= \frac{1}{m\mu}\frac{1 + 2\rho + \ldots + n\rho^{n-1}}{1 + \rho + \ldots + \rho^{n-1}}.$$

9. Call center with waiting and impatience

Assume that there are m agents. Denote by λ the arrival rate, and by μ and ν the parameters of the exponential distributions of the call durations and patience durations, respectively. The process $X(t)$ describing the number of calls has the following transition graph:

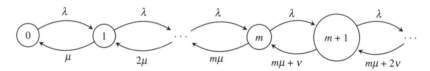

This is a birth–death process. The system is always stable and its stationary distribution is given by:

$$\pi(x) = \pi(0)\frac{\alpha^x}{x!}, \quad \text{if } x \leq m$$

$$\pi(x) = \pi(0)\frac{\alpha^m}{m!}\frac{\lambda^{x-m}}{(m\mu + \nu)\ldots(m\mu + (x-m)\nu)}, \quad \text{otherwise,}$$

where $\alpha = \lambda/\mu$ denotes the traffic intensity in erlangs. According to the PASTA property, the waiting probability is

$Q = \sum_{x \geq m} \pi(x)$. The rate of abandonment is the ratio of the frequencies of the corresponding events (see section 5.11), that is, $\lambda^{-1} \sum_{x > m} \pi(x)(x - m)\nu$.

10. Video on demand

The system corresponds to a multiclass Erlang model. We apply formula [8.10] with $\alpha_1 = 10$, $\alpha_2 = 3$, $c_1 = 1$, $c_2 = 5$, and $m = 50$. We get:

$$B_1 \approx 1.98 \times 10^{-3} \quad \text{and} \quad B_2 \approx 1.50 \times 10^{-2}.$$

11. The generalized Engset model

In the absence of blocking, the system state defines a Markov process on $\{0, 1\}^K$ with stationary measure:

$$\pi(x) = \pi(0) \prod_{k=1}^{K} \beta_k^{x_k}.$$

In the presence of admission control, the stationary distribution of the system state follows from the truncation and normalization of this measure on the state space $\mathcal{X} = \{x : |x| \leq m\}$, with $|x| = x_1 + \ldots + x_K$.

By the formula of conditional transitions (see section 5.11), the blocking probability of user 1 is the ratio of the frequency of user 1 blocking to the frequency of user 1 call arrivals, that is:

$$B_1 = \frac{\sum_{x:x_1=0,|x|=m} \nu_1 \pi(x)}{\sum_{x:x_1=0,|x|\leq m} \nu_1 \pi(x)} = \frac{\sum_{x:|x|=m} \prod_{k\neq 1} \beta_k^{x_k}}{\sum_{x:|x|=m} \prod_{k\neq 1} \beta_k^{x_k}}.$$

This is the probability that all circuits are occupied in a virtual system *without* user 1. In particular, the blocking probability of user 1 is independent of her traffic intensity.

The probability that the m circuits are occupied is given by:

$$p = \frac{\beta_1 \sum_{x:|x|=m-1} \prod_{k\neq 1} \beta_k^{x_k} + \sum_{x:|x|=m} \prod_{k\neq 1} \beta_k^{x_k}}{\beta_1 \sum_{x:|x|\leq m-1} \prod_{k\neq 1} \beta_k^{x_k} + \sum_{x:|x|\leq m} \prod_{k\neq 1} \beta_k^{x_k}}.$$

Since the blocking probability of any user decreases with m, this is the case of user 1:

$$\frac{\sum_{x:|x|=m-1} \prod_{k\neq 1} \beta_k^{x_k}}{\sum_{x:|x|\leq m-1} \prod_{k\neq 1} \beta_k^{x_k}} \geq \frac{\sum_{x:|x|=m} \prod_{k\neq 1} \beta_k^{x_k}}{\sum_{x:|x|\leq m} \prod_{k\neq 1} \beta_k^{x_k}},$$

so that p increases with β_1. We deduce that the probability that the m circuits are occupied is an increasing function of $\beta_1, \ldots, \beta_K$. In particular, the blocking probability of a user is maximized when the traffic intensities of the other users is maximized. We conclude that the user who experiences the highest blocking probability is that with the lowest traffic intensity, namely user $k^\star$ with $k^\star = \arg\min_k \beta_k$.

Chapter 9

Real-time Traffic

The following two chapters are dedicated to IP networks whose resources are not reserved but dynamically shared by active users: this is the "best-effort" paradigm of the Internet. In case of congestion, incoming flows are still admitted (there is no admission control), and thus contribute to the quality deterioration of all active flows. It is then essential to be able to estimate the probability of congestion events and to dimension the network so as to keep this probability as low as possible.

This chapter is dedicated to real-time flows such as voice, video conference, and audio or video streaming; the next chapter is dedicated to data transfers.

9.1. Flows and packets

As the name suggests, real-time traffic is characterized by a time constraint that does not allow the source to react to congestion. It is typically transferred by the user datagram protocol (UDP). For the sake of simplicity, flows are assumed to have a constant bit rate (CBR) (see exercise 6 in section 9.11 for an extension to variable bite rate (VBR) flows). Figure 9.1 illustrates the traffic fluctuations at both flow and packet levels when all flows have the same bit rate, in this case 10 packets/s.

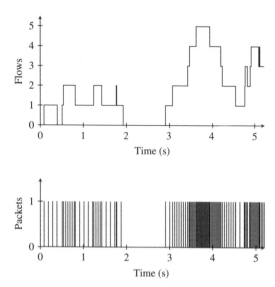

Figure 9.1. *Real-time traffic at flow (top) and packet (bottom) levels*

When the number of flows is constant, traffic is periodic at the packet level. In practice, packets are delayed in routers on their path to the destination, which adds some randomness in the process. In fact, this randomness is negligible compared with that due to the variation of the number of active flows, as shown below.

9.2. Packet-level model

First, we consider the traffic fluctuations at packet level on short time scales, with a fixed number of active flows.

While packets arrive, in theory, according to a periodic process, in practice, this process is disturbed by the random queuing delays. Therefore, we assume that the packets arrive according to a Poisson process of intensity λ. This can be considered as a conservative assumption (see section 3.8). For the sake of simplicity, packets are also assumed to have an exponential size of mean σ bits, which is also a conservative assumption (in practice, packet sizes are bounded). Denote by $A = \lambda\sigma$ the traffic intensity (in $\mathrm{bit/s}$) and by C the link capacity (in $\mathrm{bit/s}$).

The transmission time of a packet having an exponential distribution with parameter $\mu = C/\sigma$, the model corresponds to an $M/M/1$ queue with arrival rate λ and service rate μ. The load of this queue is given by:

$$\rho = \frac{\lambda}{\mu} = \frac{A}{C}.$$

The queue is stable if and only if $\rho < 1$. Under this assumption, the distribution of the number of packets x in the queue is given by:

$$\forall x \in \mathbb{N}, \quad \pi(x) = (1 - \rho)\rho^x.$$

According to the PASTA property, the probability that a packet finds at least x packets in the queue at its arrival is given by:

$$\sum_{y \geq x} \pi(y) = \rho^x.$$

This probability decreases exponentially fast with x. An example is given in Figure 9.2 for $x = 25, 50$, or 100 packets (corresponding to a mean delay $x\sigma/C$ of 2, 4, or $8\,\mathrm{ms}$ for $\sigma = 1\,\mathrm{kB}$ and $C = 100\,\mathrm{Mbit/s}$).

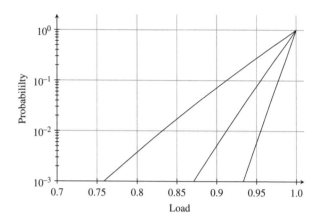

Figure 9.2. *Probability that the number of packets is larger than x*
(x = 25, 50, 100, from top to bottom)

Note that, for $x = 100$, this probability is negligible when the load is smaller than 0.9. In fact, this is negligible for loads very close to 1 with more realistic assumptions on the traffic (less random packet arrivals and bounded packet sizes).

If the size of the buffer is also considered, the model becomes an $M/M/1/n$ queue, where n denotes the maximum number of packets in the router, including the packet in transmission. The distribution of the number of packets in the queue is then:

$$\forall x = 0, 1, \ldots, n, \quad \pi(x) = \frac{\rho^x}{1 + \rho + \ldots + \rho^n}.$$

The packet loss probability follows from the PASTA property:

$$\pi(n) = \frac{\rho^n}{1 + \rho + \ldots + \rho^n}.$$

As illustrated in Figure 9.3, this quantity is very small when $\rho < 1$ and close to the fraction of traffic in excess $(\rho - 1)/\rho$ when $\rho > 1$, for realistic values of the buffer size (at least 100 packets).

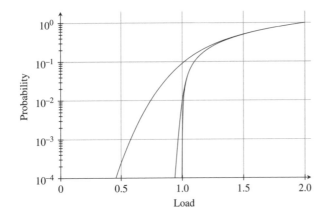

Figure 9.3. *Packet loss probability with respect to the load ρ for different values of the buffer size (n = 10, 100, 1,000, from top to bottom)*

9.3. Flow-level model

Since the traffic randomness at packet level is not likely to generate significant packet delays and loss rates, the rest of the chapter is dedicated to the study of the traffic fluctuations at flow level. We assume that flows arrive according to a Poisson process of intensity λ and have exponential durations with parameter μ (these notations should not be confused with those used in section 9.2). As for the Erlang model, the results are in fact insensitive to the distribution of flow durations beyond the mean (see exercise 2 in section 9.11). The traffic intensity in erlangs is given by:

$$\alpha = \frac{\lambda}{\mu}.$$

A typical performance metric of interest is the congestion rate, that is the probability that the total rate of flows is larger than the link capacity, as illustrated in Figure 9.4.

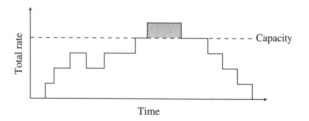

Figure 9.4. *A congestion event*

Since the duration of each flow is independent of the state of the system, the flow-level model corresponds to an $M/M/\infty$ queue. Thus, the number of flows X has a Poisson distribution with mean α:

$$\forall x \in \mathbb{N}, \quad \pi(x) = e^{-\alpha}\frac{\alpha^x}{x!}. \tag{9.1}$$

First, assume that all flows have the same CBR r and share the same link of capacity C, with $C = mr$ for some integer m. Each flow has a mean volume of r/μ, so that the traffic intensity (in bit/s) is equal to:

$$A = \alpha r.$$

The notations are summarized in Table 9.1.

The link load is the ratio of traffic intensity to link capacity:

$$\rho = \frac{A}{C} = \frac{\alpha}{m}.$$

Note that the system is stable at any load because the duration of each flow is independent of the congestion state of the network; of course, the link load has an impact on the distribution of the number of ongoing flows, see equation [9.1], and thus on the frequency of congestion events.

	Absolute values (bit/s)	Relative values (multiples of r)
Traffic	A	α
Capacity	C	m

Table 9.1. *Main notations*

9.4. Congestion rate

The congestion rate is defined as the probability that flows are constrained by the link:

$$G = \mathrm{P}(X > m),$$

which follows from equation [9.1]. Figure 9.5 shows the congestion rate with respect to the link load ρ for different values of capacity. Observe that the congestion rate decreases with capacity at low loads, whereas it increases with capacity at high loads. As for the Erlang formula, this can be explained by the lower relative traffic fluctuations for large α (see exercise 1 in section 8.9): there is almost no congestion when $\alpha < m$ and almost always congestion when $\alpha > m$. When $m \to \infty$, the congestion rate is null if $\rho < 1$ and equal to 1 otherwise.

Note that an active flow sees a higher congestion rate, equal to:

$$G' = \mathrm{P}(X \geq m),$$

since it is enough that m other flows are active for the link to be saturated (see exercise 1 in section 9.11). This is the congestion rate for a link capacity $C - r$. The difference with the congestion rate G is significant only for small values of capacity.

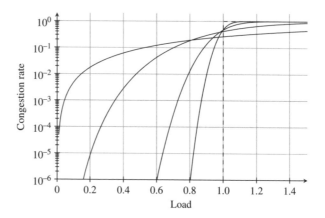

Figure 9.5. *Congestion rate (m = 1, 10, 100, 500, ∞, from top to bottom at low load)*

9.5. Mean throughput

The congestion rate is the probability that the flow throughput is less than the required CBR r. Many other throughput metrics can be calculated, such as the mean, the variance, or some quantiles of the flow throughput, in order to estimate the quality deterioration due to the network. For this purpose, we need to model the system in congestion events. Assume that the link capacity is *fairly* shared between flows, so that each flow has throughput C/x when $xr > C$. Each flow is represented as a "fluid" stream of information with a constant rate r going through a bottleneck of capacity C, the traffic in excess being lost. Thus, in the presence of x flows, each flow gets a throughput equal to:

$$\min\left(\frac{C}{x}, r\right).$$

From the point of view of an active flow, the distribution of the number of flows (including itself) is equal to the initial distribution weighted by the number of flows (see section 6.10), that is:

$$\pi'(x) \propto x\pi(x).$$

Consider, for instance, the mean throughput, which we denote by γ. We have:

$$
\begin{aligned}
\gamma &= \sum_{x \geq 1} \min\left(\frac{C}{x}, r\right) \pi'(x) \\
&= \frac{\sum_{x \geq 1} \min(\frac{C}{x}, r) x\pi(x)}{\sum_x x\pi(x)} \\
&= r\frac{\mathrm{E}(\min(X, m))}{\mathrm{E}(X)}.
\end{aligned}
\tag{9.2}
$$

Using equation [9.1], we get:

$$
\begin{aligned}
\mathrm{E}(\min(X, m)) &= \mathrm{E}(X1(X \leq m)) + m\mathrm{P}(X > m) \\
&= \sum_{x=1}^{m} e^{-\alpha} \frac{\alpha^x}{(x-1)!} + mG \\
&= \alpha(1 - B)(1 - G) + mG,
\end{aligned}
$$

where G is the congestion rate and B the corresponding Erlang formula for link capacity m and traffic intensity α. Using the fact that $\mathrm{E}(X) = \alpha$, we deduce:

$$\gamma = r(1 - B)(1 - G) + \frac{r}{\rho}G. \tag{9.3}$$

Figure 9.6 shows the mean throughput γ with respect to the link load ρ for different capacity values and a unit CBR $r = 1$. Note that, at constant load ρ, the mean throughput increases with capacity. When $m \to \infty$, the mean throughput is equal to 1 if $\rho < 1$ (since both G and B are null) and to $1/\rho$ otherwise (since $G = 1$).

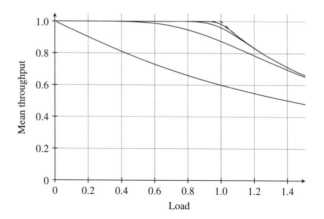

Figure 9.6. *Mean throughput (m = 1, 10, 100, 500, ∞,
from bottom to top)*

9.6. Loss rate

Another useful performance metric is the packet loss rate. In the presence of x active flows, and under the previous assumption of "fluid" traffic, the traffic lost per time unit is null if $xr \leq C$ and equal to $xr - C$ otherwise. Since the system spends a fraction $\pi(x)$ of time in state x, we deduce the loss rate:

$$L = \frac{\sum_{x:xr>C}(xr - C)\pi(x)}{\sum_x xr\pi(x)}$$

$$= \frac{\mathrm{E}((X - m)\,1(X > m))}{\mathrm{E}(X)}.$$ [9.4]

Writing:

$$\mathrm{E}((X - m)\,1(X > m)) = \mathrm{E}(X) - \mathrm{E}(\min(X, m)),$$

we get the following relationship from equation [9.2]:

$$L = 1 - \frac{\gamma}{r}.$$ [9.5]

The packet loss rate simply be interpreted as the mean throughput loss, relative to the required bit rate r. Thus, a mean throughput equal to 95% of the required bit rate corresponds to a packet loss rate of 5%.

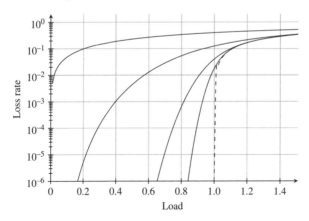

Figure 9.7. *Loss rate (m = 1, 10, 100, 500, ∞, from top to bottom)*

Figure 9.7 shows the loss rate with respect to the link load ρ for different values of capacity. Again, performance improves with capacity at constant load. When $m \to \infty$, the loss rate is null if $\rho < 1$ and equal to the fraction of traffic in excess $(\rho - 1)/\rho$ otherwise (see section 8.2).

REMARK 9.1.– (Adaptative flows). The packet loss rate is lower if flows adapt their transmission rate to congestion. Traffic is not truly real time in this case; the case of rate adaptation is considered in the next chapter.

9.7. Multirate model

In practice, flows have different characteristics, which depend on the type of service (voice, video) and the coding scheme. Consider N classes of flows. Each class i flow has a CBR of r_i bit/s. We seek to calculate the congestion rate of a link of capacity C bit/s (see exercise 5 in section 9.11 for other performance metrics). Flows are assumed to arrive according to a Poisson process. We denote by α_i the traffic intensity of class i flows (in erlangs). The corresponding traffic intensity (in bit/s) is given by:

$$A_i = \alpha_i r_i.$$

Denote by A the total traffic intensity (in bit/s):

$$A = \sum_{i=1}^{N} A_i.$$

The link load is the ratio of traffic intensity to link capacity:

$$\rho = \frac{A}{C}.$$

Let X_i be the number of class i flows. Denote by X and r the vectors $(X_1, \ldots, X_N)$ and $(r_1, \ldots, r_N)$. The system is equivalent to N independent $M/M/\infty$ queues. The stationary distribution of X is given by:

$$\forall x \in \mathbb{N}^N, \quad \pi(x) = e^{-\alpha_1} \frac{\alpha_1^{x_1}}{x_1!} \ldots e^{-\alpha_N} \frac{\alpha_N^{x_N}}{x_N!}.$$

The congestion rate is given by:

$$G = \mathrm{P}(X.r > C). \qquad [9.6]$$

As for the single rate case, a class i flow sees the congestion rate of a system with a total capacity $C - r_i$, that is:

$$G_i = \mathrm{P}(X.r > C - r_i). \qquad [9.7]$$

Results are illustrated in Figure 9.8 for $N = 4$ classes of flows as a function of load ρ, with a homogeneous traffic distribution (in bit/s), i.e. $A_1 = A_2 = A_3 = A_4$.

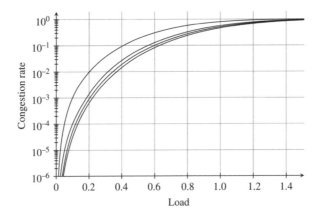

Figure 9.8. *Congestion rate seen by different classes of flows in a multirate system (C = 100, r = (1, 5, 10, 30) from bottom to top)*

9.8. Recursive formula

In order to efficiently calculate expressions equations [9.6] and [9.7], the Kaufman–Roberts formula [8.10] can be applied using normalized integer values of capacity $m = C/r_0$ and CBRs $c_1 = r_1/r_0, \ldots, c_N = r_N/r_0$, for some common bit rate unit $r_0 > 0$. This yields:

$$G = 1 - e^{-(\alpha_1 + \ldots + \alpha_N)} \sum_{n=0}^{m} f(n),$$

and:

$$G_i = 1 - e^{-(\alpha_1 + \ldots + \alpha_N)} \sum_{n=0}^{m - c_i} f(n).$$

9.9. Network models

As for circuit traffic, we can evaluate the impact of the simultaneous sharing of several resources. For the sake of simplicity, assume that all flows have some unit CBR. The other notations are those given in section 8.7. The stationary distribution of the network state X is given by:

$$\forall x \in \mathbb{N}^N, \quad \pi(x) = e^{-\alpha_1} \frac{\alpha_1^{x_1}}{x_1!} \dots e^{-\alpha_N} \frac{\alpha_N^{x_N}}{x_N!}.$$

The congestion rate as seen by class i flows is given by:

$$G_i = \mathrm{P}(X + e_i \notin \mathcal{X})$$

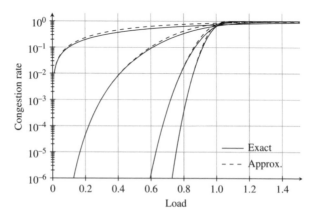

Figure 9.9. *Congestion rate of class 1 flows for the network of Figure 8.15*
($m_1 = 1, 10, 100, 250$, from top to bottom, $m_2 = 2m_1$)

Using the decoupling approximation of section 8.8, we get:

$$G_i \approx 1 - \prod_{l \in R_i} (1 - g_l),$$

where g_l denotes the congestion rate seen by an active flow on link l, considered in isolation, that is with capacity m_l (in number of flows) and traffic intensity $\sum_{i:l \in R_i} \alpha_i$. The quality of the approximation is illustrated in Figure 9.9 for the network of Figure 8.15 with the same traffic intensity for both flow classes. The same approximation can be used for the loss rate and the mean throughput (see exercise 8 in section 9.11).

9.10. Gaussian approximation

For high traffic intensities, the distribution of the number of flows (see equation [9.1]) is close to a Gaussian[1] distribution[2], as illustrated in Figure 9.10 for $\alpha = 5, 10, 25$ E. Since the mean and variance of X are both equal to α, we obtain the following approximation for large α:

$$P(X > x) \approx \frac{1}{\sqrt{2\pi\alpha}} \int_x^\infty e^{-\frac{(t-\alpha)^2}{2\alpha}} \, dt.$$

We deduce the congestion rate for a link of capacity $C = mr$:

$$G \approx \frac{1}{\sqrt{2\pi\alpha}} \int_m^\infty e^{-\frac{(t-\alpha)^2}{2\alpha}} \, dt. \tag{9.8}$$

The quality of the approximation is shown in Figure 9.11.

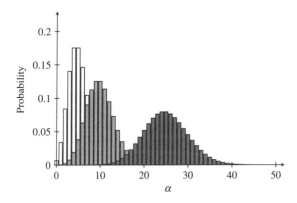

Figure 9.10. *Poisson distributions with means $\alpha = 5, 10, 25$*

---

1 Carl Friedrich Gauß, German mathematician (1777–1855).

2 This is a consequence of the central limit theorem, a Poisson variable of mean n having the same distribution as the sum of n independent Poisson variables of mean 1.

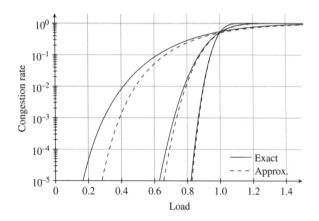

Figure 9.11. *Gaussian approximation of the congestion rate*
(m = 10, 100, 500, from top to bottom at low load)

Using the following approximation, valid for large x:

$$\int_{x}^{\infty} e^{-\frac{t^2}{2}}\, dt \approx \frac{e^{-\frac{x^2}{2}}}{x},$$

we deduce from equation [9.8] the following explicit approximation of the congestion rate for large m at load $\rho < 1$:

$$G \approx \frac{1}{\sqrt{2\pi m}} \frac{\sqrt{\rho}}{1-\rho} e^{-m\frac{(1-\rho)^2}{2\rho}}.$$

In particular, at any fixed load $\rho < 1$, the congestion rate decreases *exponentially* with the link capacity m.

The Gaussian approximation (see equation [9.8]) may also be written in terms of the capacity C and traffic intensity A expressed in bit/s. We obtain:

$$G \approx \frac{1}{\sqrt{2\pi Ar}} \int_{C}^{\infty} e^{-\frac{(t-A)^2}{2Ar}}\, dt.$$

Interestingly, this expression remains valid in the presence of multiple rates, as described in section 9.7, by defining r as the average rate weighted by the traffic intensities:

$$r = \frac{1}{A} \sum_{i=1}^{N} A_i r_i.$$

This follows from the fact that the traffic of class i has mean A_i and variance $A_i r_i$, so that the overall traffic has mean $A = \sum_{i=1}^{N} A_i$ and variance $Ar = \sum_{i=1}^{N} A_i r_i$.

9.11. Exercises

1. The observer paradox
   Using the results of section 6.10, prove that the congestion rate as seen by an active flow on a link of capacity C is equal to the congestion rate of a link of capacity $C - r$.

2. Insensitivity
   Using the same technique as in exercise 4 of section 8.9, prove that the congestion rate, the mean throughput, and the loss rate are performance metrics which depend on the distribution of flow durations through the mean only.

3. Admission control
   Consider the model of section 9.3 in which the number of active flows is limited to some value $n > m$. Which queue does this system correspond to? Give the blocking rate and the congestion rate. Find a similar formula as equation [9.3] for the mean throughput. Does relationship [9.5] still hold?

4. Silence deletion
   Consider voice-over-IP traffic using a coding scheme with unit constant rate when the source is active and a null rate otherwise (no packet is emitted). Assume that active and silence periods have exponential durations with respective parameters μ and ν. Moreover, the conversation ends after each active period with probability p; otherwise, a new active period starts after a silence period.

Calculate the mean user speaking time σ during a conversation. Assuming that the flows arrive according to a Poisson process of intensity λ, prove that the number of active flows has a Poisson distribution with parameter $\alpha = \lambda\sigma$; the system might be represented by a Jackson network of two infinite-server queues (see exercise 7 in section 7.8). What is the conclusion regarding network dimensioning?

5. *Multirate model*

Consider the multirate model of section 9.7. Assume that, in any state x, all flows have the same packet loss rate, given by $(x.r - C)/x.r$. Calculate the throughput of a class i flow in state x. Knowing that the stationary probability of state x seen by a class i active flow is proportional to $x_i\pi(x)$ (this is an extension of the results presented in section 6.10), calculate the mean throughput γ_i and the loss rate L_i of class i flows. Prove the relationship $L_i = 1 - \gamma_i/r_i$, which generalizes equation [9.5].

6. *VBR video*

Consider a coding scheme emitting packets with N different rates, denoted by $r_1, \ldots, r_N$. Assume that a video is encoded at rate r_i during an exponential duration with parameter μ_i, then encoded at rate r_j with probability p_{ij}, and ends with probability $p_i = 1 - \sum_{j=1}^{N} p_{ij}$. Each video starts at rate r_i with probability q_i. Using the technique given in section 7.1, calculate the mean number of sequences at rate r_i. Assuming that videos are generated according to a Poisson process of intensity ν, prove that the stationary distribution of the number of videos at rate r_i, for $i = 1, \ldots, N$, is the stationary distribution of a multirate model.

7. *Video server*

Consider K users sharing the same video server. Videos are transmitted in CBR on a link that can transmit up to m simultaneous flows. Video durations are exponential with parameter μ; between the end of a video and the beginning of the next one, a user remains idle for an exponential duration with parameter ν. What is the distribution of the number of simultaneous video flows? Calculate the congestion rate.

8. *Network performance*
Consider the network in Figure 8.15. Flows have a CBR r. Express the loss rate of each class of flow according to the capacities C_1 and C_2, and traffic intensities α_1 and α_2, due to the decoupling approximation introduced in section 8.8. Deduce from relation [9.5] an approximation of the mean throughput of each class of flow.

9.12. Solution to the exercises

1. *The observer paradox*
According to equation [6.14], the distribution of the number of active flows seen by an active flow is $x\pi(x)/E(X)$. Moreover, π is the stationary distribution of an $M/M/\infty$ queue. The congestion rate seen by an active flow is then given by:

$$G' = \frac{\sum_{x>m} x\pi(x)}{E(X)},$$

$$= \frac{1}{E(X)} \sum_{x>m} \frac{\alpha^x}{(x-1)!},$$

$$= \sum_{x\geq m} \pi(x).$$

This is the congestion rate of a link of capacity $(m-1)r = C-r$.

2. *Insensitivity*
The result comes from the fact that these performance metrics (congestion rate, mean throughput, and loss rate) only depend on the stationary distribution of the number of active flows, which is independent of the flow duration distribution beyond the mean.

For example, consider a hyperexponential flow duration distribution: the duration of each flow has an exponential distribution with parameter μ_1 with probability p_1 and an exponential distribution with parameter μ_2 with probability p_2, with $p_1 + p_2 = 1$ and $p_1/\mu_1 + p_2/\mu_2 = 1/\mu$. This model is equivalent to two parallel $M/M/\infty$ queues with respective arrival rates λp_1 and λp_2 and respective service rates μ_1 and μ_2. The states of these two queues are independent, reversible Markov processes with respective stationary distributions:

$$\pi_1(x_1) = e^{-\alpha_1} \frac{\alpha_1^{x_1}}{x_1!} \quad \text{and} \quad \pi_2(x_2) = e^{-\alpha_2} \frac{\alpha_2^{x_2}}{x_2!},$$

with $\alpha_1 = \lambda p_1/\mu_1$ and $\alpha_2 = \lambda p_2/\mu_2$.

The stationary distribution of the joint process is the product of the two stationary distributions:

$$\pi(x) = e^{-(\alpha_1 + \alpha_2)} \frac{\alpha_1^{x_1}}{x_1!} \frac{\alpha_2^{x_2}}{x_2!}.$$

In particular, the total number of active flows has a Poisson stationary distribution with parameter $\alpha_1 + \alpha_2 = \lambda/\mu$, as if the flow durations had an exponential distribution with parameter μ.

3. *Admission control*

This is equivalent to an $M/M/n/n$ queue with arrival rate λ and service rate μ. With $\alpha = \lambda/\mu$, its stationary distribution is given by:

$$\forall x = 1, \ldots, n, \quad \pi(x) = \frac{\frac{\alpha^x}{x!}}{1 + \alpha + \frac{\alpha^2}{2} + \cdots + \frac{\alpha^n}{n!}}.$$

The blocking rate B is given by the Erlang formula (see section 8.2) and the congestion rate is obtained as in section 9.4:

$$B = \pi(n) = \frac{\frac{\alpha^n}{n!}}{1 + \alpha + \frac{\alpha^2}{2} + \cdots + \frac{\alpha^n}{n!}},$$

$$G = \sum_{x:m<x\leq n} \pi(x) = \frac{\frac{\alpha^{m+1}}{(m+1)!} + \cdots + \frac{\alpha^n}{n!}}{1 + \alpha + \frac{\alpha^2}{2} + \cdots + \frac{\alpha^n}{n!}}.$$

Using the expression of the stationary distribution, we get mean throughput:

$$\gamma = r(1 - G') + \frac{r}{\rho}(1 - B)G.$$

where G' is the congestion rate of a network of capacity $C - r$ admitting $n - 1$ active flows. Expression [9.5] still holds.

4. Silence deletion

The number of speaking periods during a call has a geometric distribution with parameter p; the average speaking time is thus equal to $\sigma = 1/(p\mu)$.

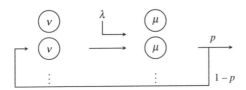

Figure 9.12. *Model of active and silent periods using a Jackson network*

Denote by $X_a(t)$ and $X_s(t)$ the number of active and silent users at time t, respectively. The system is equivalent to a Jackson network with two infinite-server queues, as represented in Figure 9.12. Results of exercise 7 in section 7.8 apply. Denote by λ_a and λ_s the solutions to the traffic equations:

$$\lambda_a = \lambda + \lambda_s \quad \text{and} \quad \lambda_s = (1 - p)\lambda_a.$$

We have $\lambda_a = \lambda/p$ and $\lambda_s = (1-p)\lambda/p$. Letting $\alpha_a = \lambda_a/\mu$ and $\alpha_s = \lambda_s/\nu$, we get the stationary distribution:

$$\pi(x_a, x_s) = e^{-(\alpha_a + \alpha_s)} \frac{\alpha_a^{x_a} \, \alpha_s^{x_s}}{x_a! \; x_s!}.$$

In particular, the number of active users has a Poisson distribution with parameter $\alpha_a = \alpha$. Thus, network dimensioning reduces to consider *actual* voice traffic, without the silence periods.

5. Multirate model

In any state x such that $x.r > C$, the throughput of a class i flow is proportional to its bit rate that is equal to:

$$\frac{r_i}{x.r} C.$$

We then get the mean throughput of a class i flow:

$$\gamma_i = r_i \frac{\sum_{x:x_i \geq 1} \min\left(\frac{C}{x.r}, 1\right) x_i \pi(x)}{\sum_{x:x_i \geq 1} x_i \pi(x)} = r_i \frac{\mathrm{E}\left(\min\left(\frac{C}{X.r}, 1\right) X_i\right)}{\mathrm{E}(X_i)}.$$

Similarly, the loss rate is given by:

$$L_i = \frac{\mathrm{E}\left(X_i \left(1 - \frac{C}{X.r}\right) 1(X.r > C)\right)}{\mathrm{E}(X_i)}.$$

Using the fact that:

$$\mathrm{E}\left(X_i \left(1 - \frac{C}{X.r}\right) 1(X.r > C)\right) = \mathrm{E}(X_i) - \mathrm{E}\left(\min\left(\frac{C}{X.r}, 1\right) X_i\right),$$

it follows that $L_i = 1 - \gamma_i/r_i$.

6. VBR video

The sequence of rates of a video is a Markov chain on the state space $\mathcal{X} = \{0, \ldots, N\}$, as introduced in section 7.2. State 0 represents the idle state; the other states represent the encoding levels. There is a transition from state 0 to state

i with probability q_i and from state i to state j with probability p_{ij}. The stationary distribution of this Markov chain is the solution to the balance equations:

$$\pi_0 = \sum_{i=1}^{N} \pi_i p_i \quad \text{and} \quad \pi_i = \pi_0 q_i + \sum_{i=1}^{N} \pi_i p_{ij}, \ i = 1, \ldots, N.$$

A video is active at encoding level i with frequency π_i and is inactive with frequency π_0. The mean number of sequences of a video at encoding level i is equal to π_i/π_0.

The active videos can be modeled by a Jackson network with infinite-server queues (see exercise 7 in section 7.8). Denote by λ_i, for $i = 1, \ldots, N$, the solutions to the traffic equations:

$$\lambda_i = \nu q_i + \sum_{j=1}^{N} \lambda_j p_{ji}.$$

Denoting by $\alpha_i = \lambda_i/\mu_i$, the stationary distribution of the number of videos at each encoding level satisfies:

$$\pi(x) = \prod_{i=1}^{N} e^{-\alpha_i} \frac{\alpha_i^{x_i}}{x_i!}, \ \forall x.$$

This is the multirate model with parameters λ_i, μ_i, and r_i, for $i = 1, \ldots, N$.

7. *Video server*

The system is equivalent to the Engset model without blocking (see section 8.3). The stationary distribution of the number of active videos π is a binomial:

$$\pi(x) = \binom{K}{x} \left(\frac{\mu}{\nu + \mu} \right)^x \left(\frac{\nu}{\nu + \mu} \right)^{K-x}, \ \forall x = 1, \ldots, K.$$

The congestion rate is given by:

$$G = \sum_{x=m+1}^{K} \pi(x).$$

8.   *Network performance*

The decoupling approximation consists of considering each link separately. Loss rates on links 1 and 2 are, respectively, given by:

$$l_1 = 1 - (1 - b_1)(1 - g_1) - \frac{g_1}{\rho_1} \quad \text{and} \quad l_2 = 1 - (1 - b_2)(1 - g_2) - \frac{g_2}{\rho_2},$$

where b_1 and g_1 (respectively, b_2 and g_2) are the blocking rate and the congestion rate for a link of capacity C_1 and traffic intensity α_1 (respectively, C_2 and $\alpha_1 + \alpha_2$), and $\rho_1 = \alpha_1/C_1$ (respectively, $\rho_2 = (\alpha_1 + \alpha_2)/C_2$).

We deduce the approximate values for the loss rates, namely $L_1 \approx 1 - (1 - l_1)(1 - l_2)$ for class 1 flows and $L_2 \approx l_2$ for class 2 flows. Using equation [9.5], we obtain the following approximations for the mean throughputs: $\gamma_1 \approx r(1 - l_1)(1 - l_2)$ and $\gamma_2 \approx r(1 - l_2)$.

Chapter 10

Elastic Traffic

Finally, we consider data transfers such as email downloads, web browsing, and peer-to-peer file sharing. Such traffic is said to be "elastic" since it adapts to the available bandwidth, the duration of data transfers depending on the obtained throughput. The results presented in this chapter are key to dimensioning IP networks, first because the majority of IP traffic is elastic, and second because, even in the presence of real-time traffic, considering all traffic as elastic is a conservative assumption. Indeed, those packets of real-time traffic that are lost are typically not retransmitted, which decrease the network load in case of congestion; for elastic traffic, all lost packets are retransmitted until they are correctly received (this is the role of the transmission control protocol, TCP) so that the whole demand is eventually processed.

10.1. Bandwidth sharing

Elastic traffic is typically controlled by TCP, which adapts the rate of sources to congestion events, mainly detected through packet losses. Each flow is thus characterized by a data *volume* and not by a *duration*, which depends on the available bandwidth. We will assume that flows arrive according to a Poisson process of intensity λ and have exponential volumes of mean σ bit. As in Chapters 8 and 9, the results are in fact insensitive to the distribution of flow sizes beyond the mean (see exercise 1 in section 10.8). We denote by $A = \lambda\sigma$ the traffic intensity in bit/s.

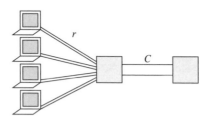

Figure 10.1. *Users with peak rate r sharing of a link of capacity C*

In the reference model, all flows have the same peak rate r. This typically corresponds to the speed of the user access line. In practice, flows share a common backhaul link of capacity C, as illustrated in Figure 10.1. We assume, for simplicity, that $C = mr$ for some integer m. Thus, the link is limiting as soon as the ongoing number of flows, x, exceeds m. The congestion control algorithms of TCP typically ensure its sharing, in an approximately fair way. In the following, we will assume that this sharing is *perfectly* fair so that each flow has rate C/x whenever $x > m$. Note that, for $m = 1$, users are constrained by the backhaul link only (since $r = C$ in this case). Like real-time traffic, it is the dynamical bandwidth sharing, illustrated in Figure 10.2, that determines network performance as seen by users.

The model corresponds to an $M/M/m$ queue under the processor-sharing service discipline, with arrival rate λ and service rate $\mu = r/\sigma$. In the absence of congestion, the mean flow duration is equal to $1/\mu$ so that the ratio $\alpha = \lambda/\mu$

corresponds to the traffic intensity expressed in erlangs; this is related to the traffic intensity in bit/s through the relationship $A = \alpha r$ (see Table 10.1 for a summary of the notations). The load of the queue is given by:

$$\rho = \frac{\lambda}{m\mu} = \frac{A}{C}.$$

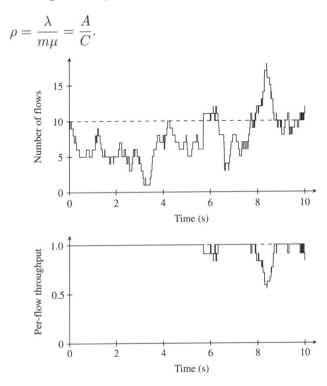

Figure 10.2. *Dynamical bandwidth sharing* $(C = 10, r = 1)$

From section 6.4, the stationary distribution of the number of ongoing flows X is given by:

$$\begin{aligned}
\pi(x) &= \pi(0)\frac{\alpha^x}{x!} &&\text{if } x \leq m \\
\pi(x) &= \pi(m)\rho^{x-m} &&\text{otherwise,}
\end{aligned}$$

[10.1]

under the stability condition $\rho < 1$. When $\rho \geq 1$, the system is unstable in the sense that the number of flows tends to grow continuously, the throughput of each flow tending to zero. In practice, it is then the behavior of users, impatient of not having received their data, that leads the system to an equilibrium point, with a very low throughput per user (see exercise 7 in section 10.8).

The notations are summarized in Table 10.1.

	Absolute values (bit/s)	Relative values (multiples of r)
Traffic	A	α
Capacity	C	m

Table 10.1. *Main notation*

In the rest of the chapter, we assume that the stability condition $\rho < 1$ is satisfied. The normalization then gives:

$$\pi(0) = \frac{1}{1 + \alpha + \frac{\alpha^2}{2} + \ldots + \frac{\alpha^m}{m!} + \frac{\alpha^m}{m!}\frac{\rho}{1-\rho}}.$$

We start with the calculation of the performance metrics introduced in Chapter 9 for real-time traffic (congestion rate, mean throughput, and loss rate), and then present some extensions to this reference model.

10.2. Congestion rate

The congestion rate is defined as the probability that the throughput of flows is constrained by the link, that is:

$$G = \mathrm{P}(X > m).$$

From equation [10.1], we obtain:

$$G = \pi(m)\frac{\rho}{1 - \rho},$$

that is:

$$G = \frac{\rho B}{1 - \rho + \rho B}. \qquad [10.2]$$

where B denotes the associated Erlang loss formula [8.3] for capacity m and traffic intensity α. From equation [8.7], we also have $G = \rho Q$, where Q denotes the associated Erlang waiting formula. Figure 10.3 illustrates the congestion rate with respect to the link load ρ for various values of capacity.

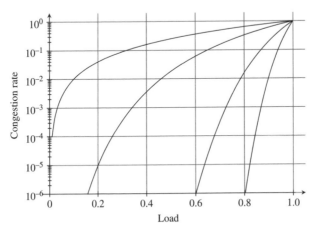

Figure 10.3. *Congestion rate with respect to load (m = 1, 10, 100, 500, from top to bottom)*

Like real-time traffic, the congestion rate *seen by users* is biased by their activity, this bias being negligible for links of high capacity (see exercise 3 in section 10.8).

10.3. Mean throughput

Under the assumption of fair sharing, each flow gets in state x a rate equal to:

$$\min\left(\frac{C}{x}, r\right).$$

The distribution of the number of ongoing flows seen by an active flow (including itself) is equal to the initial distribution π weighted by the number of flows (see section 6.10), that is,

$\pi'(x) \propto x\pi(x)$. We deduce the mean throughput of an active flow:

$$\gamma = \sum_{x \geq 1} \min\left(\frac{C}{x}, r\right) \pi'(x) = \frac{\sum_{x \geq 1} \min(\frac{C}{x}, r)x\pi(x)}{\sum_x x\pi(x)}$$

$$= r\frac{E(\min(X, m))}{E(X)}. \tag{10.3}$$

By the conservation law (see section 6.5), the traffic at the input of the link must be equal to the traffic at the output of the link. Since the output rate is equal to $\min(xr, C)$ in state x, we obtain:

$$A = rE(\min(X, m)).$$

Thus, we have:

$$\gamma = \frac{A}{E(X)}. \tag{10.4}$$

Using equation [10.1], we finally get:

$$\gamma = r\frac{\rho(1 - \rho)m}{G + \rho(1 - \rho)m}. \tag{10.5}$$

REMARK 10.1.– (Mean flow duration). From equation [10.4] and Little's formula, the mean throughput can also be interpreted as the ratio of the mean flow size to the mean flow duration.

Figure 10.4 illustrates the mean throughput γ with respect to the link load ρ for various values of capacity and a unit peak rate, $r = 1$. Notice that for $m = 1$, the mean throughput is equal to $1 - \rho$, and thus linear in the load; for $m = 500$, the mean throughput is very close to the peak rate, even at high load.

A simple bound can be deduced from the inequality $G \leq \rho$ (recall that $G = \rho Q$, where Q is the associated Erlang waiting formula) so that:

$$\gamma \geq r \frac{(1 - \rho)m}{1 + (1 - \rho)m}.$$ [10.6]

If $m = 100$, for instance, the ratio of mean throughput to peak rate γ/r is higher than 0.9 whenever the load ρ is less than 0.9. This bound shows the scale economies: the mean throughput tends to the peak rate when capacity m grows, at constant load.

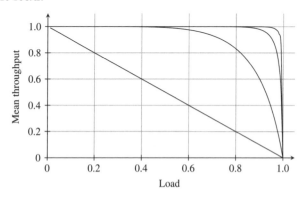

Figure 10.4. *Mean throughput with respect to load ($m = 1, 10, 100, 500$, from bottom to top)*

10.4. Loss rate

Like real-time traffic, we assume that the total rate of flows at the input of the link is equal to xr in the presence of x flows; thus, we neglect the adaptive nature of flows due to TCP that tends to decrease the loss rate. Under this conservative assumption, we obtain as in section 9.6:

$$\begin{aligned} L &= \frac{\sum_{x:xr>C} (xr - C)\pi(x)}{\sum_x xr\pi(x)} \\ &= \frac{\mathrm{E}((X - m)\,1(X > m))}{\mathrm{E}(X)}. \end{aligned}$$ [10.7]

Again, writing:

$$\mathrm{E}((X - m)\,\mathbb{1}(X > m)) = \mathrm{E}(X) - \mathrm{E}(\min(X, m)),$$

we deduce from equation [10.3] the following relationship between the loss rate and the mean throughput:

$$L = 1 - \frac{\gamma}{r}. \qquad [10.8]$$

From equation [10.5], we get:

$$L = \frac{G}{G + \rho(1 - \rho)m}. \qquad [10.9]$$

The results are illustrated in Figure 10.5. We can again use the inequality $G \leq \rho$ to obtain the following simple bound:

$$L \leq \frac{1}{1 + (1 - \rho)m}. \qquad [10.10]$$

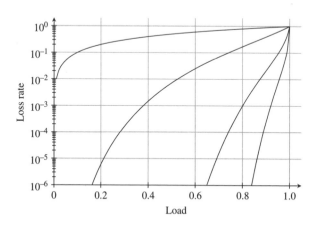

Figure 10.5. *Loss rate with respect to load (m = 1, 10, 100, 500, from bottom to top)*

10.5. Multirate model

Finally, we present some extensions to the reference model studied so far. We start with the multirate model. Specifically, we consider N classes of flows, flows of class i having peak rate r_i bit/s. Class i flows arrive according to a Poisson process of intensity λ_i and have exponential volumes with mean σ_i (the results are, in fact, insensitive to this distribution beyond the mean). The traffic intensity of class i flows is equal, in bit/s, to:

$$A_i = \lambda_i \sigma_i.$$

In the absence of congestion, class i flows have exponential durations with parameter $\mu_i = r_i/\sigma_i$. We denote by $\alpha_i = \lambda_i/\mu_i$ the associated traffic intensity in erlangs, related to the traffic intensity in bit/s through the relationship $A_i = \alpha_i r_i$. The total traffic intensity, in bit/s, is given by:

$$A = \sum_{i=1}^{N} A_i.$$

We seek to calculate the performance metrics induced by the fair sharing of a link of capacity C bit/s. The link load is given by:

$$\rho = \frac{A}{C}.$$

Let x_i be the number of class i flows. We denote by x and r the vectors $(x_1, \ldots, x_N)$ and $(r_1, \ldots, r_N)$. The system corresponds to N $M/M/.$ queues whose service rates depend on the global system state x. Let $m_i(x)$ be the total rate of class i flows, expressed in peak rate units r_i. The service rate of the corresponding queue is equal to $\mu_i m_i(x)$ in state x, with:

$$m_i(x) = x_i \qquad \text{if } x.r \leq C$$

$$\sum_{i=1}^{N} r_i m_i(x) = C \quad \text{otherwise.} \qquad [10.11]$$

The first equality states that each flow gets its peak rate in the absence of congestion; the second states that the total rate of flows is equal to the link capacity in case of congestion (the considered bandwidth sharing is assumed perfectly efficient).

To make the model tractable, we seek a sharing policy for which these queues form a Whittle network. We define the function $\Phi(x)$ by:

$$\Phi(x) = \begin{cases} \dfrac{1}{x_1! \ldots x_N!} & \text{if } x.r \leq C \\[2em] \dfrac{1}{C} \displaystyle\sum_{i:x_i \geq 1} r_i \Phi(x - e_i) & \text{otherwise.} \end{cases} \qquad [10.12]$$

We verify from equation [7.15] that the associated service rates satisfy the constraints [10.11] as well as the balance condition [7.13] of Whittle networks. In particular, we deduce from equation [7.14] the stationary distribution of the numbers of flows of each class:

$$\pi(x) = \begin{cases} \pi(0) \dfrac{\alpha_1^{x_1} \ldots \alpha_N^{x_N}}{x_1! \ldots x_N!} & \text{if } x.r \leq C \\[2em] \dfrac{1}{C} \displaystyle\sum_{i:x_i \geq 1} A_i \pi(x - e_i) & \text{otherwise.} \end{cases} \qquad [10.13]$$

The various performance metrics follow as in the reference model. In particular, the congestion rate is the probability that $x.r > C$; the congestion rate *seen by class i flows* is given by:

$$G_i = \frac{\sum_{x:x.r>C} x_i \pi(x)}{\sum_x x_i \pi(x)} = \frac{\mathrm{E}(X_i 1(X.r > C))}{\mathrm{E}(X_i)}. \qquad [10.14]$$

Figure 10.6 illustrates the corresponding results with respect to the link load ρ for $N = 4$, $C = 100$, $r = (1, 5, 10, 30)$, and a homogeneous traffic distribution in bit/s, that is, with $A_1 = A_2 = A_3 = A_4$.

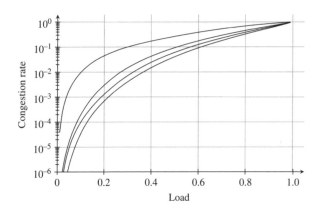

Figure 10.6. *Congestion rate seen by various classes of flows in a multirate system (C = 100, r = (1, 5, 10, 30) from bottom to top)*

The total rate of class i flows being equal to $m_i(x)r_i$ in state x, the mean throughput of class i flows is given by:

$$\gamma_i = \frac{\sum_{x:x_i \geq 1} \frac{m_i(x)r_i}{x_i} x_i \pi(x)}{\sum_{x:x_i \geq 1} x_i \pi(x)} = \frac{\mathrm{E}(m_i(X)r_i)}{\mathrm{E}(X_i)}, \qquad [10.15]$$

that is, by the conservation law:

$$\gamma_i = \frac{A_i}{\mathrm{E}(X_i)}. \qquad [10.16]$$

Assuming that each flow transmits at its peak rate at the input of the link, we obtain the loss rate of class i flows:

$$L_i = \frac{\sum_x (x_i - m_i(x))r_i \pi(x)}{\sum_x x_i r_i \pi(x)} = \frac{\mathrm{E}(X_i - m_i(X))}{\mathrm{E}(X_i)}.$$

Using equation [10.15], we deduce the relationship:

$$L_i = 1 - \frac{\gamma_i}{r_i}. \qquad [10.17]$$

10.6. Recursive formula

The Kaufman–Roberts formula [8.10] can be extended to elastic traffic using normalized integer values of capacity $m = C/r_0$ and peak rates $c_1 = r_1/r_0, \ldots, c_N = r_N/r_0$, for some common bit rate unit $r_0 > 0$. Let:

$$f(n) = \sum_{x:x.c=n} \frac{\alpha_1^{x_1}}{x_1!} \cdots \frac{\alpha_N^{x_N}}{x_N!},$$

for all $n = 0, 1, \ldots, m$, and:

$$\bar{f} = \sum_{x:x.c>m} \frac{\pi(x)}{\pi(0)}.$$

Note that the congestion rate is given by:

$$G = \frac{\bar{f}}{f(0) + f(1) + \ldots + f(m) + \bar{f}}.$$

The values of $f(1), \ldots, f(m)$ follow from the Kaufman–Roberts formula [8.10]. Moreover, we have from equation [10.13]:

$$\bar{f} = \sum_{x:x.c>m} \frac{\pi(x)}{\pi(0)}$$

$$= \sum_{x:x.c>m} \sum_{i=1}^{N} \frac{A_i}{C} \frac{\pi(x - e_i)}{\pi(0)}$$

$$= \sum_{i=1}^{N} \frac{A_i}{C} \sum_{x:x.c>m-c_i} \frac{\pi(x)}{\pi(0)}$$

$$= \sum_{i=1}^{N} \frac{A_i}{C} \left(\bar{f} + \sum_{n=m-c_i+1}^{m} f(n) \right),$$

with the convention that $\pi(x) = 0$ if $x \notin \mathbb{N}^N$. We deduce:

$$\bar{f} = \sum_{i=1}^{N} \frac{A_i \bar{f}_i}{C - A}, \qquad\qquad [10.18]$$

with:

$$\bar{f}_i = \sum_{n=m-c_i+1}^{m} f(n).$$

Regarding the network state seen by class i flows, let:

$$g_i(n) = \sum_{x:x.c=n} x_i \frac{\alpha_1^{x_1}}{x_1!} \cdots \frac{\alpha_N^{x_N}}{x_N!},$$

for all $n = 1,\ldots,m$, and:

$$\bar{g}_i = \sum_{x:x.c>m} \frac{x_i \pi(x)}{\pi(0)}.$$

Note that the congestion rate [10.14] of class i flows is given by:

$$G_i = \frac{\bar{g}_i}{g_i(0) + g_i(1) + \ldots + g_i(m) + \bar{g}_i}, \qquad [10.19]$$

while their mean throughput [10.16] is given by:

$$\gamma_i = A_i \frac{f(0) + f(1) + \ldots + f(m) + \bar{f}}{g_i(0) + g_i(1) + \ldots + g_i(m) + \bar{g}_i}. \qquad [10.20]$$

We have:

$$g_i(n) = \sum_{x:x.c=n,x_i\geq 1} \frac{\alpha_1^{x_1}}{(x_1-1)!} \cdots \frac{\alpha_N^{x_N}}{x_N!} = \alpha_i f(n-c_i),$$

for all $n = 1, \ldots, m$, and, in view of [10.13]:

$$
\begin{aligned}
\bar{g}_i &= \sum_{x:x.c>m} x_i \frac{\pi(x)}{\pi(0)} \\
&= \sum_{x:x.c>m} x_i \sum_{j=1}^{N} \frac{A_j}{C} \frac{\pi(x - e_j)}{\pi(0)} \\
&= \sum_{j=1}^{N} \frac{A_j}{C} \sum_{x:x.c>m} x_i \frac{\pi(x - e_j)}{\pi(0)} \\
&= \sum_{j=1}^{N} \frac{A_j}{C} \left(\bar{g}_i + \sum_{n=m-c_j+1}^{m} g_i(n) \right) + \frac{A_i}{C}(\bar{f} + \bar{f}_i).
\end{aligned}
$$

We deduce:

$$
\bar{g}_i = \frac{A_i}{C - A}(\bar{f} + \bar{f}_i) + \sum_{j=1}^{N} \frac{A_j \bar{g}_{ij}}{C - A}, \qquad [10.21]
$$

with:

$$
\bar{g}_{ij} = \sum_{n=m-c_j+1}^{m} g_i(n).
$$

10.7. Network model

As for real-time traffic, we can estimate the impact of the simultaneous sharing of multiple resource and use the decoupling approximation to calculate the resulting performance metrics whatever the network size is. The quality of the approximation is shown in Figure 10.7 for the congestion rate seen by class 1 flows in the network of Figure 8.15, with the same traffic intensity for both classes of flows. The exact value is calculated for a sharing policy leading to a Whittle network, such as the multirate model.

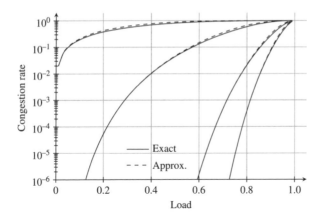

Figure 10.7. *Congestion rate of class 1 flows in the network of Figure 8.15*
($C_1 = 1, 10, 100, 250$, from top to bottom, $C_2 = 2C_1$)

10.8. Exercises

1. Insensitivity
  Show, as in exercise 4 of section 8.9, that the congestion rate, the mean throughput, and the loss rate are insensitive to the flow size distribution beyond the mean.

2. Congestion rate
  Show using equation [10.2] that the congestion rate is less than or equal to ρ^2. In which case there will be equality?

3. Congestion rate seen by users
  Show using the results of section 6.10 that the congestion rate as seen by users is given by:

$$G' = \frac{\mathrm{E}(X\,1(X > m))}{\mathrm{E}(X)}.$$

Applying the conservation law, show that:

$$\mathrm{E}(X1(X \leq m)) = (\rho - G)m.$$

Deduce the expression:

$$G' = G\,\frac{1 + (1 - \rho)m}{G + \rho(1 - \rho)m}.$$

Calculate the ratio G'/G for $m = 1$ and $m \to \infty$, at constant load ρ.

4. IP link dimensioning

We consider an IP link of $150\,\text{Mbit/s}$ connecting K users. Calculate the maximal value of K for a target congestion rate of 1%, given that each user generates $100\,\text{kbit/s}$ on average and has a peak rate of $10\,\text{Mbit/s}$. We assume that all flows are elastic and arrive according to a Poisson process.

5. Admission control

We consider the model of section 10.1 in which the ongoing number of flows is limited to some maximum value $n > m$. What is the corresponding queue? Give the blocking rate of flows, as well as their mean throughput. Show that this mean throughput is equal to the ratio of the mean flow size to the mean flow duration (see Remark 10.1).

6. Access link

We consider K users sharing an access link to the Internet. Each user generates a sequence of elastic flows of exponential sizes of mean σ; between the end of a flow and the beginning of the next flow, the user stays idle during an exponential duration with parameter ν. We assume that flows are constrained only by the access link, of capacity C. Calculate the congestion rate and the mean throughput.

7. Impatience

We consider the reference model in which each user may abandon their data transfer if it does not complete quickly enough. Specifically, the patience of each user has an exponential duration with parameter ν. Give the stationary measure of the Markov process $X(t)$ describing the number of ongoing flows at time t. Give the stability condition. Calculate the fraction of abandoned flows and the mean throughput.

8. *2G radio access*

The GPRS and EDGE technologies use the timeslots initially dedicated to voice for data transfers. Each mobile can use up to four simultaneous slots. We consider eight slots dedicated to GPRS. In the presence of more than two mobiles, these slots are evenly shared by the mobiles. Calculate the mean throughput of a mobile for a traffic intensity of $40\,\mathrm{kbit/s}$. Each slot is assumed to provide a data rate of $10\,\mathrm{kbit/s}$.

9. *3G+ radio access*

The 3G+ technology is based on the time sharing of the radio resource. Under round-robin scheduling, this sharing is perfectly fair. Thus, in the presence of n mobiles in the cell, the throughput of each mobile is equal to its physical rate divided by n. We consider two typical locations in the cell: at location 1, the physical rate is equal to $C_1 = 1\,\mathrm{Mbit/s}$; at location 2, the physical rate is equal to $C_2 = 2\,\mathrm{Mbit/s}$. Elastic flows arrive at locations 1 and 2 according to Poisson processes with respective intensities λ_1 and λ_2; the flow sizes are exponential with mean σ.

Show that the system corresponds to a Whittle network (see section 7.6). Deduce the stationary distribution. We define the load ρ as the fraction of time where the base station is active. Calculate ρ and show that the stability condition is given by $\rho < 1$. Calculate the mean throughput of flows at locations 1 and 2 with respect to ρ. Give the values obtained for $\lambda_1 = \lambda_2 = 0.5$ flows per second and $\sigma = 100\,\mathrm{kB}$.

10.9. Solution to the exercises

1. Insensitivity

The result follows from the fact that the various performance metrics (congestion rate, mean throughput, and loss rate) depend on the distribution of the number of ongoing flows, which is independent of the flow size distribution beyond the mean.

Like in exercises 4 and 2 of respective Chapters 8 and 9, we consider a hyperexponential flow size distribution: the size of each flow has an exponential distribution with mean σ_1 with probability p_1 and an exponential distribution with mean σ_2 with probability p_2, with $p_1 + p_2 = 1$ and $p_1\sigma_1 + p_2\sigma_2 = \sigma$. The model corresponds to two $M/M/\cdot$ queues in parallel with respective arrival rates $\lambda p_1, \lambda p_2$ and service rates $\mu_1 m_1(x), \mu_2 m_2(x)$, with $\mu_1 = r/\sigma_1$, $\mu_2 = r/\sigma_2$ and:

$$m_1(x) = x_1 \qquad m_2(x) = x_2 \qquad \text{if } x_1 + x_2 \leq m$$

$$m_1(x) = \frac{x_1}{x_1 + x_2}m \quad m_2(x) = \frac{x_2}{x_1 + x_2}m \quad \text{otherwise.}$$

These queues are coupled through their service rates. We verify that the balance equation [7.13] is satisfied so that the system corresponds to a Whittle network. From equation [7.14], the associated stationary measure is given by:

$$\pi(x) = \pi(0)\Phi(x)\alpha_1^{x_1}\alpha_2^{x_2},$$

with $\alpha_1 = \lambda_1/\mu_1$, $\alpha_2 = \lambda_2/\mu_2$ and:

$$\Phi(x) = \frac{1}{x_1!x_2!} \qquad\qquad \text{if } x_1 + x_2 \leq m$$

$$\Phi(x) = \binom{x_1 + x_2}{x_1}\frac{1}{m!m^{x_1+x_2-m}} \qquad \text{otherwise.}$$

In particular, the stationary distribution of the total number of active flows is given by equation [10.1], with $\alpha = \alpha_1 + \alpha_2$ and $\rho = \alpha/m$, and thus coincides with that obtained for an exponential flow size distribution with mean σ.

2. *Congestion rate*

Expression [10.2] is increasing in B at constant load ρ. Since B decreases with capacity at constant load (see section 8.2), B and G are maximum when capacity m is equal to 1. In this case, $B = \rho/(1 + \rho)$ and $G = \rho^2$.

3. *Congestion rate seen by users*

Let π' be the distribution of the number of users as seen by an observer in steady state. In view of section 6.10, we have $\pi'(x) \propto x\pi(x)$. Moreover, $G' = \sum_{x>m} \pi'(x)$. After normalization, we get:

$$G' = \frac{\sum_{x>m} x\pi(x)}{\mathrm{E}(X)} = \frac{\mathrm{E}(X\mathbf{1}(X > m))}{\mathrm{E}(X)}.$$

Applying conservation law [6.7], we obtain:

$$\sum_{x \leq m} x\pi(x) + m \sum_{x>m} \pi(x) = \rho m.$$

We deduce that $\mathrm{E}(X\mathbf{1}(X \leq m)) = (\rho - G)m$. From the stationary distribution [10.1], we have:

$$\mathrm{E}(X) = \frac{G + \rho(1 - \rho)m}{(1 - \rho)},$$

so that:

$$G' = G\frac{1 + (1 - \rho)m}{G + \rho(1 - \rho)m}.$$

In particular, $G'/G = (2-\rho)/\rho$ for $m = 1$ and $G'/G \to 1/\rho$ when $m \to \infty$.

4. IP link dimensioning

This is an application of the reference model with traffic intensity $A = K \times 100\,\text{kbit/s}$, capacity $C = 150\,\text{Mbit/s}$, and ratio capacity to peak rate $m = 15$. From equation [10.2], we can connect at most 795 users for a target congestion rate of 1%; the associated load is then equal to 53%.

5. Admission control

This is an $M/M/m/n$ queue. The stationary distribution is given by:

$$\pi(x) = \pi(0)\frac{\alpha^x}{x!}, \text{ for } x = 0, 1, \ldots, m$$

$$\pi(x) = \pi(m)\rho^{x-m}, \text{ for } x = m, \ldots, n.$$

From the PASTA property, the blocking rate is $B = \pi(n)$. The mean throughput of an admitted flow is given by:

$$\gamma = \frac{\sum_{x=1}^{n} \min(\frac{C}{x}, r)x\pi(x)}{\sum_{x=1}^{n} x\pi(x)} = r\frac{\text{E}(\min(X, m))}{\text{E}(X)}.$$

From the conservation law [see equation [6.8]], the admitted traffic is equal to the traffic at the output of the link so that $A(1 - B) = r\text{E}(\min(X, m))$. We obtain:

$$\gamma = \frac{A(1 - B)}{\text{E}(X)}.$$

The arrival rate of admitted flows being equal to $\lambda(1 - B)$, it follows from Little's formula that the mean throughput is the ratio of the mean flow size to the mean flow duration.

6. Access link

Let $X(t)$ be the number of active users at time t. This is a Markov process whose transition graph is the following, with $\mu = C/\sigma$:

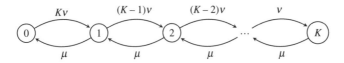

This is a birth–death process, whose stationary distribution is given by:

$$\pi(x) = \pi(0)\frac{\beta^x}{(K-x)!}, \quad \forall x = 1, \ldots, K,$$

with $\beta = \nu/\mu$. Flows do not get their peak rate as soon as at least two users are active; thus, the congestion rate is given by $G = 1 - \pi(0) - \pi(1)$, that is:

$$G = \frac{K(K-1)\beta^2 + \cdots + K!\beta^K}{1 + K\beta + K(K-1)\beta^2 + \cdots + K!\beta^K}.$$

As in section 6.10, we obtain the mean throughput:

$$\gamma = \frac{\sum_{x=1}^{K}\frac{C}{x}x\pi(x)}{\sum_{x=1}^{K}x\pi(x)} = C\frac{1-\pi(0)}{\mathrm{E}(X)},$$

that is:

$$\gamma = \frac{1 + (K-1)\beta + \cdots + (K-1)!\beta^{K-1}}{1 + 2(K-1)\beta + 3(K-1)(K-2)\beta^2 + \ldots + K!\beta^{K-1}}.$$

7. *Impatience*

We obtain the following transition graph:

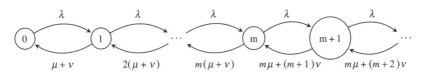

This is a birth–death process, whose stationary measure is given by:

$$\pi(x) = \begin{cases} \pi(0)\dfrac{1}{x!}\left(\dfrac{\lambda}{\mu+\nu}\right)^{x} & \text{if } x \leq m \\[4mm] \pi(m)\displaystyle\prod_{k=m+1}^{x}\dfrac{\lambda}{m\mu+k\nu} & \text{otherwise.} \end{cases}$$

The measure π has a finite sum so the Markov process $X(t)$ is stable. We now denote by π the stationary distribution. In view of 5.11, the fraction of abandoned flows is given by:

$$f = \frac{\sum_{x\geq 1}\pi(x)x\nu}{\sum_{x\geq 1}\pi(x)\mu(x)},$$

with:

$$\mu(x) = \min(x,m)\mu + x\nu.$$

By the conservation law, we obtain:

$$f = \frac{\nu}{\lambda}\mathrm{E}(X).$$

The mean throughput of flows is given by:

$$\gamma = r\frac{\mathrm{E}(\min(X,m))}{\mathrm{E}(X)}.$$

By the conservation law, we have:

$$\mu\mathrm{E}(\min(X,m)) = \lambda(1-f) \quad \text{and} \quad \nu\mathrm{E}(X) = \lambda f,$$

so that:

$$\gamma = r\frac{\nu(1-f)}{\mu f}.$$

8.   2G radio access

This is a direct application of the reference model with $C = 80\,\text{kbit/s}$, $r = 40\,\text{kbit/s}$, $m = 2$, and $\alpha = 1$. We deduce that $\rho = 1/2$. In this case, the Erlang formula gives $B = 1/5$. Applying equation [10.2], we obtain $G = 1/6$. Finally, applying equation [10.5], we get $\gamma = 10\,\text{kbit/s}$.

9.   3G+ radio access

We denote by x the vector of the numbers of mobiles in each location of the cell. We define the following service capacities, for all $x \neq 0$:

$$m_1(x) = C_1 \frac{x_1}{x_1 + x_2},$$

$$m_2(x) = C_2 \frac{x_2}{x_1 + x_2}.$$

These satisfy the balance equations [7.13] so that the network is a Whittle network. We define $\rho_1 = \lambda_1 \sigma / C_1$, $\rho_2 = \lambda_2 \sigma / C_2$, and $\rho = \rho_1 + \rho_2$. Any stationary measure π satisfies:

$$\pi(x) \propto \binom{x_1 + x_2}{x_1} \rho_1^{x_1} \rho_2^{x_2}.$$

Summing over states x such that $x_1 + x_2$ is constant, we verify that the system is stable if and only if $\rho < 1$. The stationary distribution π is then given by:

$$\pi(x) = (1 - \rho) \binom{x_1 + x_2}{x_1} \rho_1^{x_1} \rho_2^{x_2}.$$

In particular, the total number of users has a geometric distribution with parameter $1 - \rho$ and the fraction of time where the base station is active is indeed equal to ρ.

As in section 6.10, the stationary distribution as seen by a mobile at location i satisfies $\pi'(x) \propto x_i \pi(x)$. We deduce:

$$\gamma_1 = C_1 \frac{\sum_x \pi(x) \frac{x_1}{x_1 + x_2}}{E(X_1)},$$

$$= C_1(1 - \rho)\rho_1 \frac{\sum_{x_1 \neq 0} \binom{x_1 + x_2 - 1}{x_1 - 1} \rho_1^{x_1} \rho_2^{x_2}}{E(X_1)},$$

$$= \frac{C_1 \rho_1}{E(X_1)}.$$

Moreover:

$$E(X_1) = \frac{\rho_1}{1 - \rho},$$

so that $\gamma_1 = (1 - \rho)C_1$. Similarly, we have $\gamma_2 = (1 - \rho)C_2$. With the proposed values, we have $\rho_1 = 0.4$ and $\rho_2 = 0.2$, that is, $\rho = 0.6$, $\gamma_1 = 0.4\,\mathrm{Mbit/s}$, and $\gamma_2 = 0.8\,\mathrm{Mbit/s}$.

Chapter 11

Network Performance

Finally, we present some applications of the previous results to engineering issues in networks and computer systems. These examples illustrate both the strength and the limits of the presented tools, some problems having no simple analytical solution. We note the diversity of the used models that are variations of the generic models introduced in the previous chapters.

11.1. IP access networks

We first consider an IP access link of capacity C bit/s. We seek the maximum number of users K that can be connected to the Internet through this link, given some target mean throughput per user, denoted by $\gamma^\star$. Each user has the same peak rate r, determined by the service provider. We assume that $C = mr$ for some integer m. We denote by a the mean traffic intensity per user, defined as the traffic she would generate in the absence of the considered access link. The total traffic intensity is $A = Ka$.

For numerical applications, we consider a link of capacity $C = 100\,\text{Mbit/s}$. The peak rate is $r = 1\,\text{Mbit/s}$ and the traffic intensity per user $a = 200\,\text{kbit/s}$. We set the target mean throughput at either $\gamma^\star = 400\,\text{kbit/s}$ or $900\,\text{kbit/s}$.

We assume elastic traffic, with perfectly fair sharing between the ongoing flows. The link load is the ratio of the offered traffic intensity to capacity, that is:

$$\rho = \frac{A}{C}.$$

11.1.1. *Poisson arrivals*

A first approach consists of assuming Poisson flow arrivals. We must then have $\rho < 1$ for the system to be stable. From equation [10.5], the mean throughput per user is given by:

$$\gamma = r\frac{\rho(1 - \rho)m}{G + \rho(1 - \rho)m},$$

where G denotes the associated congestion rate. This mean throughput is close to the peak rate even at high load as soon as m is sufficiently large ($m \geq 100$, for instance), as illustrated in Figure 10.4. From equation [10.6], we have in fact:

$$\gamma \geq r\frac{(1 - \rho)m}{1 + (1 - \rho)m},$$

so that $\gamma \geq \gamma^\star$ whenever:

$$\rho \leq 1 - \frac{\gamma^\star}{m(r - \gamma^\star)}.$$

For the proposed numerical values, the corresponding maximum load is close to 1. Thus, the maximum number of users is essentially determined by the stability limit $\rho < 1$, that is:

$$K < \frac{C}{a} = 500$$

11.1.2. *Finite population*

Another approach consists of using an Engset-type model, with each user having active and idle periods. This model gives more accurate results on the maximum number of users. Like the Engset model (see section 8.3), the arrival rate of new flows decreases with congestion, since a larger number of users are already active in this case: performance thus improves and a larger number of users can be connected for the same target mean throughput per user.

We assume, for simplicity, that each activity period corresponds to the transfer of a single data flow. The idle periods have exponential durations[1] with parameter ν and the data-flows have exponential sizes[1] with mean σ. This is the model of exercise 6 in section 10.8, accounting for the peak rate.

Let $\mu = r/\sigma$ and $\beta = \nu/\mu$. For a mean duration of $1/\nu$ between the end of a transfer and the beginning of the next transfer, the traffic generated by each user is given by:

$$a = \frac{\sigma}{\frac{\sigma}{r} + \frac{1}{\nu}} = r\frac{\beta}{\beta + 1} \qquad [11.1]$$

The number of active users can be described by a Markov process whose transition graph is given in Figure 11.1. We obtain the following stationary distribution:

$$\pi(x) = \begin{cases} \pi(0)\dbinom{K}{x}\beta^x & \text{if } x \leq m \\[3mm] \pi(m)\dfrac{(K-m)!}{(K-x)!}\left(\dfrac{\beta}{m}\right)^{x-m} & \text{otherwise.} \end{cases}$$

---

1 By the insensitivity property, the results depend, in fact, on these random variables through their means only, see Chapter 10.

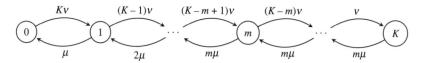

Figure 11.1. *Transition graph of the process describing the number of active users*

From equation [10.3], the mean throughput is given by:

$$\gamma = r\frac{\mathrm{E}(\min(X, m))}{\mathrm{E}(X)}.$$

For target throughputs of $\gamma^\star = 400$ and $900\,\mathrm{kbit/s}$, the number of users K cannot exceed 650 and 500, respectively.

11.1.3. *Approximation*

When the target mean throughput is relatively low compared with the peak rate, we can in fact assume that the system is in permanent congestion and that the actual traffic is close to the capacity. Since the mean flow duration is equal to σ/γ (see section 10.3), the actual traffic per user is given by:

$$b = \frac{\sigma}{\frac{\sigma}{D} + \frac{1}{\nu}} = \frac{\beta}{\frac{\beta}{D} + \frac{1}{r}}.$$

From equation [11.1], we obtain the following simple relationship between offered traffic a, actual traffic b, peak rate r, and mean throughput γ:

$$\frac{1}{a} + \frac{1}{\gamma} = \frac{1}{b} + \frac{1}{r}. \tag{11.2}$$

For the proposed numerical values, and using the approximation $C \approx Kb$, we obtain a maximum number of users equal to 650 and 511 for target mean throughputs of $\gamma^\star = 400$ and $900\,\mathrm{kbit/s}$, respectively. We conclude that the approximation is accurate in the first case due to the low target mean throughput.

11.2. 2G mobile networks

Consider a base station of a 2G mobile network serving users for both voice (Global System for Mobile Communications, GSM) and data (GPRS or EDGE). This technology relies on some time-frequency sharing of the radio resource: each base station is allocated a fixed number of frequency subcarriers, each being divided into eight slots, as illustrated in Figure 11.2.

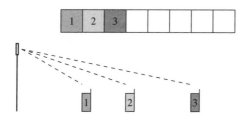

Figure 11.2. *Time sharing of a frequency subcarrier in a 2G cell*

We assume, for simplicity, that each slot is dedicated to either voice or data, which allows us to analyze each type of traffic separately, using the corresponding circuit and elastic traffic models described in Chapters 8 and 10, respectively. In fact, some slots can be dynamically allocated to voice or data; the analysis of the behavior of such a system requires a more complex model combining circuit traffic and elastic traffic (see exercise 5 in section 11.8).

11.2.1. *Voice traffic*

Let C be the radio capacity dedicated to voice in the number of slots. We assume that calls arrive according to a Poisson process of intensity λ and have exponential durations with parameter μ; the latter assumption is actually not necessary due to the insensitivity property, see Chapter 8. We neglect the impact of handovers (see exercise 2 in section 11.8 for an aspect of this problem). We denote by $\alpha = \lambda/\mu$ the traffic intensity in erlangs.

When each call requires one slot, the base station can handle at most C simultaneous calls. The call blocking rate is then given by the Erlang formula [8.3] for link capacity $m = C$ and traffic intensity α. Thus, for $10\,\mathrm{E}$ at the busy hour (i.e. 10 simultaneous calls on average), we need $C = 17$ slots for a target blocking rate of 2%.

Now consider an adaptive coding technique allowing the base station to handle two simultaneous calls on the same slot through time sharing. This increases the cell capacity at the expense of the quality of voice communications. We assume for simplicity that all mobiles are compatible with this coding technique[2] and can switch from the nominal mode referred to as *full rate* to the degraded mode referred to as *half rate* during the communication. To preserve the quality of calls, the communications are switched to the degraded mode only when required (i.e. when all the slots are occupied at the arrival of a new call).

The call blocking rate is now given by the Erlang formula [8.3] for link capacity $m = 2C$ and traffic intensity α. For $10\,\mathrm{E}$ at the busy hour, with a target blocking rate of 2%, we need $C = 9$ slots. The congestion rate, defined as the fraction of time in which the quality of some communications is degraded, is given by:

$$G = \sum_{x=C+1}^{2C} \pi(x),$$

---

2 Exercise 3 in section 11.8 considers the case where this technique is not compatible with all mobiles.

with:

$$\pi(x) = \frac{\frac{\alpha^x}{x!}}{1 + \alpha + \frac{\alpha^2}{2} + \ldots + \frac{\alpha^{2C}}{(2C)!}}, \qquad x = 0, 1, \ldots, 2C.$$

The congestion rate G' seen by an active user follows from the weighting of the stationary distribution by the number of ongoing communications. Since $2(x - C)$ communications are degraded in state x, we obtain:

$$G' = \frac{\sum_{x=C+1}^{2C} 2(x - C)\pi(x)}{\sum_{x=1}^{2C} x\pi(x)}.$$

For $C = 9$ slots, we get $G = 54\%$ and $G' = 35\%$ for an offered traffic of $10\,\mathrm{E}$, $G = 3\%$ and $G' = 2\%$ for $5\,\mathrm{E}$. Note that, in this case, the congestion rate seen by an active user is smaller than the standard congestion rate. This is due to the fact that the congestion affects only a subset of the communications.

11.2.2. *Data traffic*

Now consider the data traffic. The bursty nature of traffic leads 3GPP to normalize GPRS and EDGE technologies, enabling a much more flexible use of the radio slots than in GSM. First, each mobile can use several slots simultaneously (up to four slots on the downlink, and two on the uplink). Second, the mobiles can share the radio slots in case of congestion (up to eight mobiles per slot, typically). Here, we focus on the downlink; the approach is similar for the uplink.

Let C be the radio capacity dedicated to data, expressed in the number of slots. We assume, for convenience, that $C = 4m$, where m is a positive integer. Users generate data flows according to a Poisson process of intensity λ; flow sizes are exponential with mean σ bits. We denote by $A = \lambda\sigma$ the traffic intensity in $\mathrm{bit/s}$.

Let x be the number of ongoing data flows. As long as $x \leq m$, each data flow uses four slots simultaneously and get some throughput r (the EDGE technology allows us to adapt this throughput to the radio conditions of the mobile; see exercise 4 in section 11.8). When $x > m$, the flows share the available slots, each flow having throughput $(m/x) \times r$. We limit the number of mobiles per slot to two, so that the number of flows x cannot exceed $n = 8m$; when this limit is reached, new flows are blocked and lost.

We seek to dimension the radio capacity (i.e. to find m) for an offered traffic $A = 500\,\text{kbit/s}$, given some peak rate $r = 300\,\text{kbit/s}$, a target mean throughput $\gamma^{\star} = 100\,\text{kbit/s}$ and target blocking rate $B^{\star} = 2\%$. The model corresponds to that described in section 10.1, with admission control (see exercise 5 of Chapter 10). Specifically, the system is an $M/M/m/n$ queue of load $\rho = A/(mr)$ under the processor sharing service discipline.

Denoting by $\alpha = A/r$ the traffic intensity in erlangs, we obtain the stationary distribution of the number of ongoing flows:

$$\pi(x) = \begin{cases} \pi(0)\frac{\alpha^x}{x!} & \text{for } x = 0, 1, \ldots, m \\ \pi(m)\rho^{x-m} & \text{for } x = m, \ldots, n. \end{cases}$$

From the PASTA property, the blocking rate is $B = \pi(n)$. The mean throughput of an admitted flow is given by:

$$\gamma = \frac{\sum_{x=1}^{n} \min(r, \frac{m}{x}r)x\pi(x)}{\sum_{x=1}^{n} x\pi(x)} = r\frac{\text{E}(\min(X, m))}{\text{E}(X)}.$$

From the conservation law [6.8], the admitted traffic is equal to the output traffic, so that:

$$A(1 - B) = r\text{E}(\min(X, m)).$$

We deduce:

$$\gamma = \frac{A(1-B)}{E(X)}.$$

For the proposed numerical values, the throughput constraint is limiting. The required radio capacity is $m = 2$, that is, $C = 8$ slots. For this value, the actual mean throughput is $\gamma = 110\,\text{kbit/s}$ and the blocking rate is $B \approx 1\%$.

11.3. 3G mobile networks

3G mobile networks use a code division multiple access that is described in the Universal Mobile Telecommunications System (UMTS) norm, as illustrated in Figure 11.3. Each user requests a virtual circuit with a constant throughput, adapted to the service (voice, data, streaming). A power control algorithm is used by the transmitter to ensure a sufficient signal-to-noise ratio for the reception. The uplink and downlink are analyzed independently since radio resources are typically separated due to a frequency-division duplex.

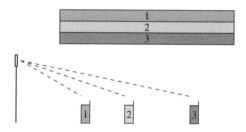

Figure 11.3. *Code-based radio resource sharing in a 3G cell*

11.3.1. *Uplink*

In the uplink, the transmission power is controlled by the mobile. Each mobile has to cope with thermal noise and interferences due to mobiles in its cell and in the neighboring cells.

First, a single type of service is considered, voice, for instance. The target signal-to-noise ratio[3] that guarantees sufficiently good communication quality is denoted by $\gamma^\star$. Let x be the number of active mobiles in the cell. Each mobile is subject to an intracell interference[4] equal to $P - P/x$, where P denotes the total signal power received by the base station. Intercell interference is usually represented as a constant fraction f of the total received power P. The signal-to-noise ratio is then given by:

$$\gamma = \frac{P/x}{N_0 + P(1 - 1/x + f)},$$

where N_0 denotes power of thermal noise. In a dense network where the thermal noise power N_0 can be neglected with respect to the received power P, the maximum number of active mobiles that can satisfy the constraint $\gamma \geq \gamma^\star$ is given by:

$$m = \left\lfloor \frac{1 + 1/\gamma^\star}{1 + f} \right\rfloor.$$

Thus, for a target signal-to-noise ratio $\gamma^\star = -18\,\text{dB}$ and an intercell interference factor $f = 0.6$, the uplink capacity of the cell is $m = 40$ calls. According to the Erlang formula, for a target blocking rate of 2%, the cell sustains a traffic up to $30\,\text{E}$.

This method can be extended to a set of N services, characterized by their target signal-to-noise ratios $\gamma_1^\star, \ldots, \gamma_N^\star$. Let $x_1, \ldots, x_N$ be the numbers of active mobiles in the cell for each type of service. If P denotes the total power received by the base station and r_i the fraction of power dedicated to a

---

3 Noise here refers to all the useless signals, that is, thermal noise *and* interference.

4 As opposed to the downlink, intracell communications are asynchronous and thus totally interfering.

mobile of service i, the signal-to-noise ratio for a mobile of service i becomes:

$$\gamma_i = \frac{r_i P}{N_0 + P(1 - r_i + f)}.$$

Neglecting the thermal noise, the constraint $\gamma_i \geq \gamma_i^\star$ is satisfied if and only if $r_i \geq r_i^\star$, with:

$$r_i^\star = \frac{1 + f}{1 + 1/\gamma_i^\star}.$$

The constraint is then satisfied for all services if and only if:

$$\sum_{i=1}^{N} x_i r_i^\star \leq 1.$$

The system corresponds to the multiclass Erlang model described in section 8.5 with a unit link capacity corresponding to the power resource. The blocking rates can be calculated accordingly.

Consider, for instance, $N = 2$ services with a target signal-to-noise ratio $\gamma_1^\star = -18\,\mathrm{dB}$ and $\gamma_2^\star = -12\,\mathrm{dB}$. For an intercell interference factor $f = 0.6$, we get $r_1^\star \approx 0.025$ and $r_2^\star \approx 0.1$. If service 1 represents 90% of the traffic in erlangs, the cell can sustain a total traffic of approximately $17.5\,\mathrm{E}$ with a target blocking rate of 2% for each service. Since service 2 requires four times more resources than service 1, it is the blocking rate of service 2 that is limiting; the blocking rate of service 1 for this maximum traffic load is much less than 2%.

11.3.2. *Downlink*

In the downlink, the transmission power toward each mobile is controlled by the base station. We denote by P the total transmission power. As for the uplink, we first consider a single service that is characterized by a target signal-to-noise ratio $\gamma^\star$. The cell is divided into N areas with homogeneous radio conditions.

Denote by g_i the radio channel gain of the base station to a mobile in area i and by f_i the intercell interference factor, given by the sum of the radio channel gains of the different base stations interfering with the mobile. Denoting by r_i the fraction of total power dedicated to each mobile in area i and by β the intracell interference factor ($\beta = 0$ if mobiles in the same cell do not interfere with each other), we get the signal-to-noise ratio for each mobile of area i:

$$\gamma_i = \frac{r_i P g_i}{N_0 + P(\beta g_i(1 - r_i) + f_i)}.$$

In particular, neglecting the thermal noise, the constraint $\gamma \geq \gamma^\star$ is satisfied by area i mobiles if and only if $r_i \geq r_i^\star$, with:

$$r_i^\star = \frac{\beta + f_i/g_i}{\beta + 1/\gamma^\star}.$$

The constraint is satisfied for all areas if and only if:

$$\sum_{i=1}^{N} x_i r_i^\star \leq 1,$$

where $x_1, \ldots, x_N$ denote, respectively, the numbers of active mobiles in areas $1, \ldots, N$. This model also corresponds to the multiclass Erlang model described in section 8.5. It is worth noticing that the different required power levels here reflect the heterogeneity of the radio conditions in the cell and not the different services. In practice, the ratio f_i/g_i is negligible at the cell center and is typically around 2 at the cell edge. With a target signal-to-noise ratio $\gamma^\star = -18\,\text{dB}$ and an intracell interference factor $\beta = 0.5$, we obtain a target fraction of received power $r_i^\star$ less than 0.01 for areas i at the cell center and around 0.04 at the cell edge.

The blocking rates given by formulas [8.9] can be applied if the mobiles are assumed to stay in the same area during all the communication. Consider, for instance, a regular hexagonal network with a path loss exponent equal to 3.5 (typical of an urban area) and a homogeneous traffic distribution in the cell. Dividing the cell into a sufficiently high number of areas, it follows from Kaufman-Roberts formula [8.10] that each cell can sustain a traffic of $32\,E$ with a target blocking rate of 2% at the cell edge.

The results extend to a set of M services. Denote by $\gamma_j^\star$ the target signal-to-noise ratio for service j, by x_{ij} the number of active mobiles in area i with service j, and by r_{ij} the fraction of emitting power dedicated to each mobile in area i with service j. For such a mobile, the signal-to-noise ratio becomes:

$$\gamma_{ij} = \frac{r_{ij}Pg_i}{N_0 + P(\beta g_i(1 - r_{ij}) + f_i)}.$$

Neglecting the thermal noise, the constraint $\gamma_{ij} \geq \gamma_j^\star$ is satisfied if and only if $r_{ij} \geq r_{ij}^\star$, with:

$$r_{ij}^\star = \frac{\beta + f_i/g_i}{\beta + 1/\gamma_j^\star}.$$

The constraint is then satisfied for all areas and all services if and only if:

$$\sum_{i=1}^{N} \sum_{j=1}^{M} x_{ij} r_{ij}^\star \leq 1.$$

Consider, for instance, $N = 2$ services with target signal-to-noise ratios $\gamma_1^\star = -18\,\text{dB}$ and $\gamma_2^\star = -12\,\text{dB}$. Under the previous assumptions, with service 1 representing 90% of the traffic in erlangs, we obtain a maximum traffic of approximately $19\,E$ with a target blocking rate of 2% for each service.

11.4. 3G+ mobile networks

The High Speed Downlink Packet Access (HSDPA) norm that is used on the downlink of 3G+ networks relies on time sharing of the radio resource. With a *round-robin* algorithm, this sharing is strictly fair: in the presence of x mobiles in the cell, each mobile has a throughput equal to its transmission rate (i.e. the throughput at which it receives data when served by the base station) divided by x. Data flows are assumed to be elastic and arrive according to a Poisson process of intensity λ. Their size is exponential with mean σ bits (again, this assumption is not essential). We denote by $A = \lambda\sigma$ the traffic intensity in $\mathrm{bit/s}$.

11.4.1. *Homogeneous case*

We first assume that all mobiles have the same transmission rate, denoted by r. The system then corresponds to the reference model described in section 10.1 with a link of capacity $C = r$. This reduces to an $M/M/1$ queue of load $\rho = A/C$. The system is stable if and only if $\rho < 1$. Under this condition, we deduce from equations [10.2] and [10.3] the congestion rate $G = \rho^2$ and the mean throughput of each user $\gamma = r(1 - \rho)$.

11.4.2. *Heterogeneous case*

We now consider a more realistic model in which the transmission rate of each mobile depends on its location in the cell. The cell is divided into N areas and the transmission rate is equal to r_i in area i. Each new data flow is generated by a mobile in area i with probability p_i; according to the subdivision property of a Poisson process (see section 3.6), flows arrive in area i according to a Poisson process of intensity $\lambda_i = \lambda p_i$. Each mobile is assumed to stay in the same area during data transfer.

Let x_i be the number of active flows in area i and x be the vector $(x_1, \ldots, x_N)$. Since the radio resources are fairly shared among active mobiles, the total throughput of area i flows is equal to $r_i m_i(x)$, with:

$$m_i(x) = \frac{x_i}{x_1 + \ldots + x_N}.$$

The system can be seen as a network of N queues, as illustrated in Figure 11.4. Customers arrive in queue i according to a Poisson process of intensity λ_i and are served at rate $\mu_i m_i(x)$ in state x, with $\mu_i = r_i/\sigma$. We verify that the symmetry condition [7.13] is satisfied so that the system corresponds to a Whittle network. From equation [7.14], the stationary measure is given by:

$$\pi(x) = \pi(0) \binom{x_1 + \ldots + x_N}{x_1, \ldots, x_N} \rho_1^{x_1} \ldots \rho_N^{x_N},$$

where $\rho_i = \lambda_i/\mu_i$ corresponds to the load of area i. Denoting by $\rho = \rho_1 + \ldots + \rho_N$ the cell load, the stability condition is $\rho < 1$. Since $\pi(0) = 1 - \rho$, it corresponds to the fraction of time where the base station is active.

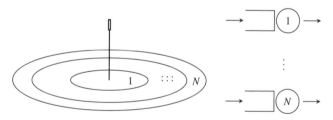

Figure 11.4. *A 3G+ cell and its representation as a Whittle network*

The stability condition defines a maximum traffic intensity that the cell can sustain, which we refer to as the *cell capacity* and denote it by C. In view of the equality:

$$\rho = A \sum_{i=1}^{N} \frac{\rho_i}{r_i},$$

we obtain:

$$C = \left(\sum_{i=1}^{N} \frac{p_i}{r_i} \right)^{-1}. \qquad [11.3]$$

Thus, the cell capacity is the *harmonic mean* of the transmission throughputs $r_1, \ldots, r_N$ weighted by the traffic distributions $p_1, \ldots, p_N$.

Under the stability condition $\rho < 1$ (equivalently, $A < C$), the total number of ongoing flows $n = x_1 + \ldots + x_N$ has the following distribution:

$$\bar{\pi}(n) = \sum_{x:x_1+\ldots+x_N=n} \pi(x) = (1-\rho)\rho^n, \quad \forall n \geq 0.$$

This is the stationary distribution of an $M/M/1$ queue with load ρ, as in the homogeneous case[5]. The congestion rate is the probability that the mobiles must share the radio resource. We obtain $G = \rho^2$ as explained previously. Moreover, the mean throughput of a mobile in area i is given by:

$$\gamma_i = \frac{\sum_{x:x_i \geq 1} \frac{r_i}{x_1+\ldots+x_N} x_i \pi(x)}{\sum_{x:x_i \geq 1} x_i \pi(x)} = \frac{A_i}{E(X_i)},$$

where $A_i = p_i A$ is the traffic intensity in area i (we used the conservation law). Since:

$$E(X_i) = \frac{\rho_i}{1-\rho},$$

we obtain:

$$\gamma_i = r_i(1-\rho),$$

---

5 This result, in fact, follows from the insensitivity property of the processor sharing service policy; see exercise 8 in section 7.8.

The mean throughput per area i user is equal to its transmission rate at load $\rho = 0$ and decreases linearly with the load.

Finally, the mean throughput in the cell is given by:

$$\gamma = \frac{\sum_{x:x_1+\ldots+x_N \geq 1} \frac{\sum_{i=1}^{N} r_i x_i}{x_1+\ldots+x_N} (x_1 + \ldots + x_N)\pi(x)}{\sum_{x:x_1+\ldots+x_N \geq 1}(x_1 + \ldots + x_N)\pi(x)}$$

$$= \frac{A}{\mathrm{E}(X_1 + \ldots + X_N)}.$$

Since:

$$\mathrm{E}(X_1 + \ldots + X_N) = \frac{\rho}{1 - \rho},$$

and $\rho = A/C$, we obtain:

$$\gamma = C(1 - \rho).$$

Thus, the mean throughput per user in the cell is equal to cell capacity at load $\rho = 0$ and decreases linearly with the load.

For instance, consider $N = 3$ areas with respective transmission rates $500\,\mathrm{kbit/s}$, $1\,\mathrm{Mbit/s}$, and $2\,\mathrm{Mbit/s}$. Assume the traffic intensity is the same in each area. In view of equation [11.3], the cell capacity is approximatively equal to $860\,\mathrm{kbit/s}$. For a load of 40%, the mean throughput per user in the cell is approximatively equal to $500\,\mathrm{kbit/s}$.

11.5. WiFi access networks

WiFi access networks rely on the IEEE 802.11 standard that defines the radio channel access mode. This mode uses the CSMA/CA algorithm, which consists of sensing the channel before transmission to avoid collisions. Since it is not possible to detect a collision during the transmission of a packet, an acknowledgment is sent by the receiver to confirm the reception. Without acknowledgment, the packet is considered lost and retransmitted after a random period whose duration increases after each collision.

The precise behavior of the IEEE 802.11 protocol is illustrated in Figure 11.5. In order to send a packet, a station:

– waits for the channel to be idle during a period of distributed interframe spacing of duration t_{DIFS};

– waits for a random number of slots chosen uniformly at random over $\{0, 1, \ldots, F - 1\}$, where F is the contention window, whenever the channel is idle;

– transmits the packet and waits for an acknowledgment, sent by the destination after a period short interframe spacing (SIFS) of duration t_{SIFS} after the reception of the packet, if the transmission is successful;

– decides that the packet is lost if no acknowledgment is received and, in this case, restarts the transmission process with a doubled contention window.

Figure 11.5. *Principle of the IEEE 802.11 protocol*

The parameters of versions b and g of the IEEE 802.11 standard are given in Table 11.1. The physical rates r_{PHY} are equal to 11 and 54 Mbit/s (in practice, the physical rate depends on the radio conditions; see exercise 6 of section 11.8). The acknowledgments have a length of 14 B and are sent at a lower physical rate to ensure the reliability of the transmission; here this rate is taken as being equal to 2 Mbit/s, which corresponds to a transmission time of $t_{\text{ACK}} = 56\,\mu\text{s}$. Each packet or acknowledgment is preceded by a physical header of duration t_{PHY}. All included, the fixed overhead per data packet induced by IEEE 802.11 is given by:

$$t_0 = t_{DIFS} + 2t_{PHY} + t_{SIFS} + t_{ACK}.$$

In addition to this fixed duration, the station delays the transmission of each packet for a random period that depends on the contention window. This window is initialized at F_{min} and doubles at each collision until it reaches the maximal value F_{max}; after seven consecutive collisions, a packet is dropped and definitively lost.

Parameter	802.11 b	802.11 g
r_{PHY}	11 Mbit/s	54 Mbit/s
t_{SLOT}	20 μs	9 μs
t_{DIFS}	50 μs	28 μs
t_{SIFS}	10 μs	10 μs
t_{PHY}	192 μs	24 μs
t_{ACK}	56 μs	56 μs
F_{min}	32	16
F_{max}	1,024	1,024

Table 11.1. *Main parameters of IEEE 802.11 b and g*

As opposed to 2G and 3G mobile networks, the uplink (to the access point) and the downlink (from the access point) share the same radio channel. First, we analyze the behavior of the system with unidirectional User Datagram Protocol (UDP) traffic only, on both the downlink and the uplink. We then analyze the impact of Transmission Control Protocol (TCP) traffic, where the TCP acknowledgments compete with the data packets for accessing the channel.

11.5.1. *UDP traffic*

In the case of unidirectional UDP traffic on the downlink, only the access point generates traffic. There is no collision and the effective throughput of the system is given by:

$$r = \frac{l}{N t_{\text{SLOT}} + t_0 + \frac{l}{r_{\text{PHY}}}},$$ [11.4]

where l denotes the packet length in bits and N the mean delay in slots induced by the contention window:

$$N = \frac{F_{\min} - 1}{2}.$$

For packets of $1{,}500\,\text{B}$, for instance, the effective throughput is approximately equal to 6 and $28\,\text{Mbit/s}$ for the versions b and g of the protocol, respectively. Note that this is slightly more than half of the corresponding physical rates.

On the uplink, the stations compete for channel access. Denote by x the number of active stations and by q the probability of collision, which is assumed to be the same for each transmission attempt for a packet. For the sake of simplicity, the impact of the maximal contention window is neglected. This implies that the number of transmission attempts of any packet has a geometric distribution with parameter q. Furthermore, the mean number of slots between two transmissions of a given station is given by:

$$N = \frac{F_{\min}}{2}(1 - q)\left(1 + 2q + (2q)^2 + \ldots\right) - \frac{1}{2}.$$

Under the condition $q < 1/2$, this number is finite. We then get:

$$N = \frac{F_{\min}}{2}\frac{1 - q}{1 - 2q} - \frac{1}{2}.$$ [11.5]

In order to compute the collision probability, we use the fact that each station transmits with probability $p = 1/(N + 1)$ when the channel is idle. Knowing that a station transmits, there is no collision if no other station transmits, which implies:

$$q = 1 - \left(1 - \frac{1}{N+1}\right)^{x-1}. \qquad [11.6]$$

Equations [11.5] and [11.6] uniquely define the collision probability q on the interval $[0, 1/2)$. Now denote by u the probability that a slot is useful (in the sense that a packet is transmitted successfully during that slot) and by v the probability that a slot is idle:

$$u = xp(1-p)^{x-1}, \quad v = (1-p)^x.$$

The mean time between two successful packet transmissions is given by:

$$\tau = v(t_{\text{SLOT}} + \tau) + (1 - u - v)\left(t_0 + \frac{l}{r_{\text{PHY}}} + \tau\right).$$

We obtain:

$$\tau = \frac{vt_{\text{SLOT}} + (1 - u - v)\left(t_0 + \frac{l}{r_{\text{PHY}}}\right)}{u},$$

and deduce the total throughput of the x UDP flows:

$$r(x) = \frac{l}{\tau(x) + t_0 + \frac{l}{r_{\text{PHY}}}},$$

where the dependence of τ on x is written explicitly. The function $r(x)$ is plotted in Figure 11.6 for both, versions b and g of the protocol.

In the limiting case $x \to \infty$, the collision probability tends to $1/2$ and, according to expression [11.6], the ratio N/x tends to $1/\ln(2)$ so that:

$$\bar{r} \equiv \lim_{x \to \infty} r(x)$$

$$= \ln(2) \frac{l}{t_{\text{SLOT}} + t_0 + \frac{l}{r_{\text{PHY}}}}.$$

This value is approximately equal to 5 and $22\,\mathrm{Mbit/s}$, respectively, for the versions b and g of the protocol.

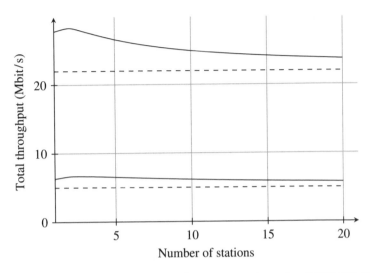

Figure 11.6. *Effective throughput of an access point using IEEE 802.11b (bottom) and IEEE 802.11g (top)*

11.5.2. *TCP traffic*

TCP traffic uses acknowledgments on the way back to regulate the data-sending rate. Consequently, the data packets compete with the TCP acknowledgments for accessing the radio channel. In the case of a single TCP flow, the station generating this flow and the access point try to access the channel simultaneously and this generates collisions.

For the sake of simplicity, assume that a TCP acknowledgment is sent for each data packet, and is denoted by l_{ACK} and l their respective lengths, in bits. Since each collision lasts for a packet transmission duration, the waiting time before transmission is given by $\tau(x)$ with $x=2$; we deduce the effective throughput of a single TCP flow:

$$r = \frac{l}{2\tau(2) + 2t_0 + \frac{l_{\text{ACK}}}{r_{\text{PHY}}} + \frac{l}{r_{\text{PHY}}}}.$$ [11.7]

For data packets of $1,500$ B and TCP acknowledgments of 40 B, this throughput is approximately equal to 4.5 and 19 Mbit/s for versions b and g of the protocol, respectively.

A notable property of TCP traffic on a WiFi access is that the total throughput is approximatively insensitive to the number of flows and to their direction of transmission (uplink or downlink). Indeed, the access point is a bottleneck at which both data packets and acknowledgments are queued. There is typically a single active station, namely the station that has just received a data packet or an acknowledgment and that sends back an acknowledgment or transmits a new data packet; the other stations wait for a packet or an acknowledgment that is queued at the access point and thus are idle. So the competition for channel access typically involves the access point and a single station, independent of the number of ongoing flows.

11.5.3. *Random traffic*

Using previous packet-level analysis, we can now account for the flow-level traffic fluctuations. In the case of TCP traffic, for instance, the WiFi access point behaves as a single link that is evenly shared by both upstream *and* downstream flows. In view of the results of Chapter 10, the mean throughput per flow on an IEEE 802.11g WiFi access with a total traffic $A = 10$ Mbit/s (both upstream and downstream, without acknowledgments) is equal to $r - A$, that is, 9 Mbit/s. With IEEE 802.11b, the access point would be overloaded and the throughput per flow would tend to zero, depending on the users' behavior (see exercise 7 of Chapter 10).

11.6. Data centers

We now consider a data center consisting of K computing units. Requests arrive according to a Poisson process of intensity λ, each requiring an exponential computing time with parameter μ. We denote by $\alpha = \lambda/\mu$ the traffic intensity in erlangs. Each unit processes the requests according to the PS service discipline and cannot accept more than m simultaneous requests. The problem is to balance the load between the K computing units, as illustrated in Figure 11.7.

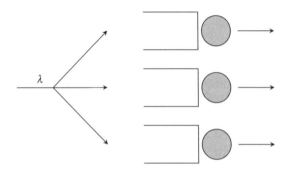

Figure 11.7. *Load balancing in a data center*

11.6.1. *Static routing*

The simplest way to share the load between the computing units is to send any incoming request to unit k with some fixed probability p_k. If the selected computing unit is processing m requests then the new request is blocked and lost. This routing is called "static" because the probabilities $p_1, \ldots, p_K$ do not depend on the system state. By the subdivision property of Poisson processes, the system behaves as K mutually independent $M/M/1/m$ queues with respective arrival rates

$\lambda p_1, \ldots, \lambda p_K$ and service rate μ. The stationary distribution of the number of requests in queue k is given by:

$$\pi_k(x_k) = \frac{\alpha_k^{x_k}}{1 + \alpha_k + \ldots + \alpha_k^m}, \quad x_k = 0, 1, \ldots, m.$$

According to the PASTA property, the blocking rate of queue k is given by $B_k = \pi_k(m)$. We deduce the mean blocking rate:

$$B = \sum_{k=1}^{K} p_k B_k = \sum_{k=1}^{K} \frac{p_k \alpha_k^m}{1 + \alpha_k + \ldots + \alpha_k^m}.$$

By symmetry, the blocking rate is minimal when the routing probabilities are uniform, that is, $p_1 = \ldots = p_K = 1/K$, in which case:

$$B = \frac{\left(\frac{\alpha}{K}\right)^m}{1 + \frac{\alpha}{K} + \ldots + \left(\frac{\alpha}{K}\right)^m}.$$

11.6.2. *Greedy routing*

A simple way to decrease the blocking rate is to send any incoming request to the least loaded computing unit. We refer to this routing scheme as "greedy". According to the PASTA property, the blocking rate is then equal to the probability that *all* computing units process m requests. This routing scheme is difficult to analyze because the associated Markov process $X(t) = (X_1(t), \ldots, X_K(t))$ describing the number of requests processed by each computing unit at time t is not reversible. For $K = 2$, for instance, there is a transition from state $(2, 0)$ to state $(1, 0)$, but not from state $(1, 0)$ to state $(2, 0)$. The computation of the blocking rates requires solving the balance equations.

11.6.3. *Adaptive routing*

We now consider an adaptive routing scheme that leads to a reversible Markov process $X(t)$ on the state space:

$$\mathcal{X} = \{x \in \mathbb{N}^K : x_1 \leq m, \ldots, x_K \leq m.\}.$$

Let $p_k(x)$ be the probability to send a request to the computing unit k when the system is in state $x \in \mathcal{X}$. This probability must be null if $x_k = m$. By simply choosing a probability $p_k(x)$ proportional to $m - x_k$, the Markov process $X(t)$ is reversible with a stationary measure:

$$\pi(x) = \pi(me) \binom{|me - x|}{me - x} \alpha^{|me-x|}, \quad x \in \mathcal{X}, \tag{11.8}$$

where e denotes the vector $(1, \ldots, 1)$ of size K and $|x|$ denotes the sum $x_1 + \ldots + x_K$. Indeed, the local balance equations:

$$\lambda p_k(x)\pi(x) = \mu\pi(x + e_k), \quad \forall x \in \mathcal{X}, x + e_k \in \mathcal{X},$$

are satisfied provided:

$$p_k(x) = \frac{m - x_k}{|me - x|},$$

which is proportional to $m - x_k$. Using the PASTA property, the blocking rate is given by:

$$B = \pi(me) = \left(\sum_{x \in \mathcal{X}} \binom{|x|}{x} \alpha^{|x|} \right)^{-1}.$$

Figure 11.8 gives the blocking rate with respect to the load α/K for $K = 2$ computing units and a maximum of $m = 10$ simultaneous requests per computing unit. We observe that the adaptive and greedy routing schemes behave similarly and outperform the static routing scheme.

11.7. Cloud computing

Finally, we consider a service of cloud computing. There are N service classes. Each class i customer receives a maximum computing capacity C_i (in flop/s[6]). The problem is to calculate the required overall computing capacity C (in flop/s), which is shared between active customers in such a way that an active class i customer gets its contracted computing capacity C_i with some probability larger than $1 - \varepsilon_i$ for some small $\varepsilon_i > 0$. For instance, the provider may offer $N = 3$ service classes, with respective computing capacities $C_1 = 10$ Gflop/s, $C_2 = 100$ Gflop/s, and $C_3 = 1$ Tflop/s, and target performance $\varepsilon_1 = 1\%$, $\varepsilon_1 = 2\%$, and $\varepsilon_3 = 5\%$.

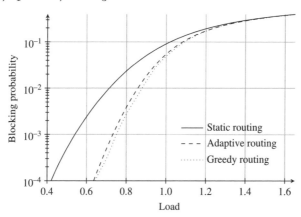

Figure 11.8. *Blocking probability with respect to the load for $K = 2$ computing units*

We assume that class i customer requests arrive according to a Poisson process of intensity λ_i and have exponential sizes (in floating operations) with mean σ_i. The corresponding traffic intensity is $A_i = \lambda_i \sigma_i$ (in flop/s). We denote by $A = \sum_{i=1}^{N} A_i$ the total traffic intensity. Assuming a fair sharing of the overall computing capacity C, the system corresponds

---

6 The acronym flop/s means *floating point operations per second.*

to the multirate model described in section 10.5, where the rates $r_1, \ldots, r_N$ are replaced by the contracted capacities $C_1, \ldots, C_N$ and the link capacity C is expressed in flop/s instead of bit/s. In particular, the congestion rate as seen by class i customers is given by equation [10.14] and can be efficiently computed through the corresponding recursive formula given by equations [10.19] and [10.21]. Figure 11.9 shows the maximum load A/C with respect to the total traffic intensity A for the $N = 3$ classes considered earlier, with homogeneous traffic distribution, i.e. $A_1 = A_2 = A_3$.

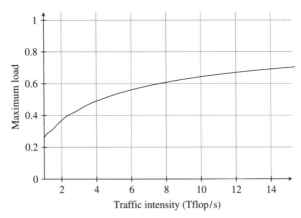

Figure 11.9. *Maximum load of a computing cloud with respect to the traffic intensity*

11.8. Exercises

1.   *Multirate IP access*

We extend the model described in section 11.1 to the case of multiple peak rates corresponding to different offers of an Internet service provider. More precisely, assume that N user classes share the same link of capacity C, class i consisting of K_i users with peak rate r_i and offered traffic a_i. Check that equation [11.2] is satisfied for each class.

Now assume that the mean throughput is a fraction η of the peak rate for each class. Give the equation satisfied by η assuming the actual total traffic is equal to the capacity. Calculate η for a link of capacity $C = 100\,\mathrm{Mbit/s}$ shared by two user classes, characterized by the following parameters: $K_1 = 500$, $K_2 = 200$, $r_1 = 1\,\mathrm{Mbit/s}$, $r_2 = 2\,\mathrm{Mbit/s}$, and $a_1 = a_2 = 200\,\mathrm{kbit/s}$.

2. Mobility and handover

A base station of a 2G mobile network has a capacity of C GSM slots. Calls arrive in the cell according to a Poisson process with intensity λ and have exponential durations with parameter μ; denote by $\alpha = \lambda/\mu$ the associated traffic intensity. A proportion p of the calls arrive from neighboring cells due to mobility: such calls are said to be in *handover*.

In order to limit the blocking rate of these calls, the following mechanism of trunk reservation is implemented: when there are n or fewer slots left, any new incoming call is blocked and lost; only calls in handover are accepted in the limit of the cell capacity. Verify that the number of ongoing calls is a birth–death process. Deduce the blocking rate of both types of calls. What is the ratio of these blocking rates when $n = 1$? Compute the corresponding value for $C = 20$ slots, $\alpha = 15\,\mathrm{E}$, and $p = 5\%$.

3. Degraded 2G calls

We are interested in the blocking rate of calls in a 2G cell when only a proportion p of the mobiles support the degraded mode (see section 11.2). Which traffic model does this system correspond to? Calculate the blocking rate of each type of mobile for $C = 12$ slots, $\alpha = 10\,\mathrm{E}$, and $p = 60\%$. During which fraction of time must a mobile supporting the degraded mode use this mode?

4.   EDGE adaptive modulation

Consider the data traffic model introduced in section 11.2, taking into account the adaptive modulation of the EDGE standard. More precisely, assume that the cell is divided into two areas in which the maximal transmission rates (on four radio slots) are equal to $r_1 = 300\,\mathrm{kbit/s}$ and $r_2 = 100\,\mathrm{kbit/s}$. Proportions p_1 and p_2 of the traffic are, respectively, in areas 1 and 2, with $p_1 + p_2 = 1$. Denote by A the offered traffic (in bit/s) and by $C = 4m$ the radio capacity in the number of slots.

Give the equivalent traffic intensity α in erlangs, that is, the mean number of occupied slots in the absence of a constraint (i.e. $C = \infty$); deduce the expression of the cell load. Compute this value for $A = 1\,\mathrm{Mbit/s}$, $C = 8$ slots, and $p_1 = p_2 = 1/2$.

Now assume that the total number of flows cannot exceed $n = 8m$. Assuming that slot sharing is perfectly fair, give the mean throughput of each type of mobile and the blocking rate.

5.   Dynamic sharing between voice and data in a 2G network

A 2G cell has C slots that are dynamically shared between voice and data. Data traffic shares the slots left by voice traffic in the limit of four slots. Specifically, data flows have access to $\min(C - x, 4)$ slots, where x is the number of voice calls. Voice and data calls arrive according to independent Poisson processes of the respective intensities λ and λ'. The other assumptions are those discussed in section 11.2.

Give the transition rates of the Markov process describing the number of voice calls and the number of data flows. Voice calls are assumed to be in a nondegraded mode and the number of data flows is limited to n. Is this process reversible? Give the mean number of slots that can be used by the data traffic.

6. *WiFi adaptive modulation*

Consider a WiFi access point shared by two users, whose radio conditions yield the respective physical rates $r_{\mathrm{PHY},1}$ and $r_{\mathrm{PHY},2}$. Calculate the total throughput r of both the users when they transfer data simultaneously using TCP.

Now assume that each user transfers a sequence of the files of exponential sizes with mean σ, with an exponential idle period with parameter ν between two data transfers. Give the transition graph of the Markov process describing the state of both the users. Is the process reversible? Give its stationary distribution, then the mean throughput of each active user. Compute the numerical values for the IEEE 802.11g standard, with $r_{\mathrm{PHY},1} = 54\,\mathrm{Mbit/s}$, $r_{\mathrm{PHY},2} = 11\,\mathrm{Mbit/s}$, $\sigma = 1\,\mathrm{MB}$, and $1/\nu = 1\,\mathrm{s}$.

11.9. Solution to the exercises

1. *Multirate IP access*

Class i users transfer files of mean size σ_i separated by idle periods of mean duration $1/\nu_i$. Let $\mu_i = r_i/\sigma_i$ and $\beta_i = \nu_i/\mu_i$. The offered traffic and the actual traffic per class i user are, respectively, given by:

$$a_i = \frac{\sigma_i}{\frac{\sigma_i}{r_i} + \frac{1}{\nu_i}} = r_i \frac{\beta_i}{\beta_i + 1} \quad \text{and} \quad b_i = \frac{\sigma_i}{\frac{\sigma_i}{\gamma_i} + \frac{1}{\nu_i}} = \frac{\beta_i}{\frac{\beta_i}{\gamma_i} + \frac{1}{r_i}},$$

where γ_i denotes the mean throughput per class i user. We deduce the relationship:

$$\frac{1}{a_i} + \frac{1}{\gamma_i} = \frac{1}{b_i} + \frac{1}{r_i}.$$

Using $\gamma_i = \eta r_i$ for a parameter $\eta < 1$, the actual total traffic is given by:

$$C \approx \sum_{i=1}^{N} K_i b_i = \sum_{i=1}^{N} \frac{K_i}{\frac{1}{a_i} + \frac{1}{r_i}\frac{1-\eta}{\eta}}.$$

For the proposed numerical values, we obtain $\eta \approx 0.5$, which corresponds to mean throughputs approximately equal to 250 and $500\,\mathrm{kbit/s}$ for class 1 and class 2 users, respectively.

2. *Mobility and handover*

The total number of ongoing calls forms a birth–death process whose transition graph is the following:

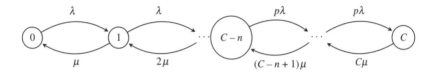

We obtain the stationary measure:

$$\pi(x) = \begin{cases} \pi(0)\frac{\alpha^x}{x!} & \text{for } x = 0, 1, \ldots, C - n \\ \pi(C - n)\frac{(p\alpha)^x}{x!} & \text{for } x = C - n + 1, \ldots, C. \end{cases}$$

The stationary distribution is obtained by normalization. From the PASTA property, the blocking rates of new calls and calls in handover are, respectively, given by:

$$B_1 = \pi(C - n) + \ldots + \pi(C) \quad \text{and} \quad B_2 = \pi(C),$$

that is:

$$B_1 = \frac{\alpha^{C-n}\left(\frac{1}{(C-n)!} + \frac{p\alpha}{(C-n+1)!} + \ldots + \frac{(p\alpha)^n}{C!}\right)}{1 + \alpha + \ldots + \alpha^{C-n}\left(\frac{1}{(C-n)!} + \frac{p\alpha}{(C-n+1)!} + \ldots + \frac{(p\alpha)^n}{C!}\right)},$$

and:

$$B_2 = \frac{\alpha^{C-n}\frac{(p\alpha)^n}{C!}}{1 + \alpha + \ldots + \alpha^{C-n}\left(\frac{1}{(C-n)!} + \frac{p\alpha}{(C-n+1)!} + \ldots + \frac{(p\alpha)^n}{C!}\right)}.$$

In particular, we have:

$$\frac{B_1}{B_2} = \frac{\frac{(p\alpha)^n}{C!}}{\frac{1}{(C-n)!} + \frac{p\alpha}{(C-n+1)!} + \cdots + \frac{(p\alpha)^n}{C!}}.$$

For $n = 1$, we get:

$$\frac{B_1}{B_2} = \frac{p\alpha}{C + p\alpha}.$$

For the proposed numerical values, the blocking rate is approximately 30 times lower for calls in handover.

3. Degraded 2G calls

We use the multiclass Erlang model discussed in section 8.5, with capacity $m = 12$ and resource requirement $c_1 = 1$ and $c_2 = 1/2$ for mobiles that do not support and support the degraded mode, respectively. The stationary distribution of the system state $x = (x_1, x_2)$ is given by:

$$\pi(x) = \pi(0) \frac{((1-p)\alpha)^{x_1}}{x_1!} \frac{(p\alpha)^{x_2}}{x_2!}, \quad x.c \le m,$$

where α denotes the total traffic intensity. The respective blocking rates are given by:

$$B_1 = \sum_{x:m-1 < x.c \le m} \pi(x) \quad \text{and} \quad B_2 = \sum_{x:m-1/2 < x.c \le m} \pi(x).$$

For the proposed numerical values, we obtain $B_1 \approx 2.5\%$ and $B_2 \approx 1.0\%$.

Each mobile that supports the degraded mode must use this mode a fraction of time equal to:

$$G_2 = \frac{\sum_{x:x.c \le m, x_1+x_2 > m} (x_1 + x_2 - m)\pi(x)}{\sum_{x:x.c \le m} x_2 \pi(x)},$$

that is $G_2 \approx 7.2\%$.

4. EDGE adaptive modulation

The maximal transmission rate per slot is equal to $r_1/4$ and $r_2/4$ in areas 1 and 2, respectively. We deduce the traffic intensity in erlangs:

$$\alpha = 4A \left(\frac{p_1}{r_1} + \frac{p_2}{r_2} \right),$$

and the cell load:

$$\rho = \frac{\alpha}{C} = \frac{A}{m} \left(\frac{p_1}{r_1} + \frac{p_2}{r_2} \right).$$

For the proposed numerical values, we obtain $\rho = 0.83$.

The system corresponds to an $M/G/m/n$ queue of load ρ under the PS service policy. By the insensitivity property, the results of section 11.2 apply. We obtain a blocking rate of $B \approx 1.3\%$ and mean throughputs of $\gamma_1 \approx 102\,\text{kbit/s}$ and $\gamma_2 \approx 34\,\text{kbit/s}$.

5. Dynamic sharing between voice and data in a 2G network

Let (x, y) be the system state, where x is the number of voice calls and y the number of data flows. The transitions rates are given by:

$$(x, y) \to (x + 1, y) : \lambda 1(x < C)$$

$$(x, y) \to (x - 1, y) : x\mu$$

$$(x, y) \to (x, y + 1) : \lambda' 1(y < n)$$

$$(x, y) \to (x, y - 1) : \frac{r}{\sigma} \min(C - x, 4) 1(y > 0).$$

The process is not reversible because there is a transition from state $(C, 1)$ to state $(C, 0)$, but not in the reverse direction (because there is no slot available for data flows).

The number of voice calls evolves independently of the number of data flows. In particular, its stationary distribution is the same as that of the number of calls in the Erlang model with link capacity C and traffic intensity $\alpha = \lambda/\mu$:

$$\forall x = 0, 1, \ldots, C, \quad \pi(x) = \frac{\frac{\alpha^x}{x!}}{1 + \alpha + \frac{\alpha^2}{2} + \ldots + \frac{\alpha^C}{C!}}.$$

We deduce the mean number of available slots for data flows:

$$\pi(C-1) + 2\pi(C-2) + 3\pi(C-3) + 4\sum_{x=0}^{C-4}\pi(x).$$

6. WiFi adaptive modulation

Each user has the same channel access probability. In view of equation [11.7], the total user throughput is given by:

$$r = \frac{l}{2\tau(2) + 2t_0 + \frac{l_{\text{ACK}}}{r_{\text{PHY}}} + \frac{l}{r_{\text{PHY}}}},$$

with:

$$\frac{l_{\text{ACK}}}{r_{\text{PHY}}} = \frac{1}{2}\left(\frac{1}{r_{\text{PHY},1}} + \frac{1}{r_{\text{PHY},2}}\right).$$

Let x_i be the activity state of user $i = 1, 2$; $x_i = 1$ if user i is active and $x_i = 0$ otherwise. The transition graph of the corresponding Markov process is given by:

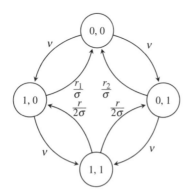

where r_i denotes the throughput of user i when alone, for $i = 1, 2$:

$$r_i = \frac{l}{2\tau(2) + 2t_0 + \frac{l_{\text{ACK}}}{r_{\text{PHY},i}} + \frac{l}{r_{\text{PHY},i}}}.$$

This process is reversible only when $r_1 = r_2$. Letting $\mu = r/\sigma$ and $\mu_i = r_i/\sigma$ for $i = 1, 2$, the balance equations are given by:

$$2\nu\pi(0,0) = \mu_1\pi(1,0) + \mu_2\pi(0,1),$$

$$(\nu + \mu_1)\pi(1,0) = \nu\pi(0,0) + \frac{\mu}{2}\pi(1,1),$$

$$(\nu + \mu_2)\pi(0,1) = \nu\pi(0,0) + \frac{\mu}{2}\pi(1,1).$$

We get:

$$\pi(1,0) = \beta\frac{\nu}{\nu + \mu_1}\pi(0,0), \quad \pi(0,1) = \beta\frac{\nu}{\nu + \mu_2}\pi(0,0),$$

$$\pi(1,1) = \frac{2\nu}{\mu}(\beta - 1)\pi(0,0),$$

with:

$$\beta = \frac{2}{\frac{\mu_1}{\nu+\mu_1} + \frac{\mu_2}{\nu+\mu_2}}.$$

We deduce the mean throughputs:

$$\gamma_1 = \pi(1,0)r_1 + \pi(1,1)\frac{r}{2} \quad \text{and} \quad \gamma_1 = \pi(0,1)r_2 + \pi(1,1)\frac{r}{2}.$$

For the proposed numerical values, we have:

$$r_1 \approx 19\,\text{Mbit/s} \quad r_2 \approx 8\,\text{Mbit/s}, \quad \text{and} \quad r \approx 11\,\text{Mbit/s}.$$

We deduce:

$$\pi(0,0) \approx 0.30, \quad \pi(1,0) \approx 0.15, \quad \pi(0,1) \approx 0.25, \quad \text{and} \quad \pi(1,1) \approx 0.30,$$

and:

$$\gamma_1 \approx 4.5\,\text{Mbit/s} \quad \text{and} \quad \gamma_2 \approx 3.7\,\text{Mbit/s}.$$

Index